DAY ON THE MOUNTAIN:

Some more writings from an author refusing to give up

AF808904

BY

A L. LOCKE

Published by Hemingway publishers

Cover design by Hemingway publishers

ISBN: Printed in United States

HEMINGWAY PUBLISHERS

Author's Note

My cigarette is smoldering in the ashtray – smoke-snake sinews trailing from the thin, white paper barrel to the heavens above.... Some unknown movie about a murdered daughter is playing in the background. It is 1:41 am, Friday morning, the 12th of January. The New Year has begun, and I, with a new book coming out, am here, down in my ground-level office, writing y'all (yes, the South has truly made an impression on me) a Love Letter to the beginning of this work.

Many of the stories included in this volume have been previously published in other books and in my online column, Notes from the Night Desk. Some of the stories are never-before-read, special pieces.

What makes this book, Days on the Mountain, so different? Well, for me, it is the short literary vignettes I am so partial to... and how I felt when I wrote them. Usually, to no surprise, it was at night when I wrote them. And I wrote them, fast and dirty and true... with little thought to content. That, sometimes, is the best way to write. Unlike my columns, interviews, and/or articles, I wrote those short pieces with a specific, furious type of love. A love I had left in a closet somewhere in the past.

Opening up that door was so, so much fun. And it was something that was needed for me, to continue on writing. In my youth, writing something profound and beautiful was very important to me. Hell, it still is, and I incorporated some of that passion in the stories within this book. I hope you, dear Reader, will enjoy them. If not, contact me and call me out, being the hack that I am.

Much love,

A. L. Locke

Table Of Contents

A Long Overdue, Thank You

What got me started writing an online blog/column? What got me started writing digital drivel for the few poor souls who call themselves my "fans?"

It was over a question.

A simple question, at that.

When I went to visit my family over Christmas vacation back in 2013… 2016…? Maybe it was in 2014 or 2017… I do not remember. Anyway, my sister asked me if I ever thought about starting a blog.

I am not sure how I responded to her, but I remember what I was thinking:

Hell, no! REAL writers do not start up blogs! Real writers write in notebooks, on typewriters, using Microsoft Word, tirelessly edit the work, rewrite the fucking thing over and over again, edit some more, and after all that love and frustration and compromise fill that mysterious void in the head and in the heart – in the pit of my stomach – you print the son of a bitch!

After it is printed, write a cover letter, write a chapter-by-chapter outline of the whole manuscript… you research for publishers and literary agents in the current edition of the Writers Market tome – you submit!

You wait for weeks.

You wait for months for an answer (oh, do not forget to send a S.A.S.E. with the already bulky package – a package that will cost $6.00 - $10.00 to send,) be sure and kiss the parcel for luck, and you do not just wait around for those envelopes to arrive.

You still have to WORK!

You still have to write while waiting for either that oh-so-sweet envelope that will make all your dreams come true, or it will be another common, disappointing thing flung from that S.A S.E.

REJECTION!

Another fucking rejection slip. More and more of them papered the walls of my workspaces across the country.

(I used to collect those pieces of paper… those physical manifestations of failure. I put a nail in more than one wall of various places I have lived, skewering those pulpy, rotten rejection slips.)

Why did I do such a thing?

I would read that Stephen King used to do the same thing. Back in his poor and youthful and desperate days when he – The KING, himself – was submitting his work with nothing but blind faith and youthful determination, he sat at a typewriter and pounded away at the keys with a vigor I am well aware of.

I have not submitted any poems or submission packets of sample chapters to any agent or publishing house in a very long time. However, I do remember riding on a plane back to Missoula, Mt., and thinking about the question my sister asked:

"Have you ever thought about starting a blog?"

I am not sure how long it has taken me to get my head out of my ass, face the situation, face the storm head-on, and start publishing my writing online.

I had to admit to myself: THIS IS WHAT WRITERS DO! They write blogs now. They publish online to build a fan base… something – anything – to get traditional publishers interested in the work.

We start blogs.

We write!

No matter what, we write.

In addition, with some persistence and a little bit of luck, someone will read what is put down.

Do I care how many people read the short rants and micro-stories I post?

Of course!

Would I still do it if they went unread?

Me posting gibberish into the darkness of cyberspace – a scream into that darkness of an electronic canyon produces a high.

Well, I'm not sure how many pages I have typed up on typewriters, how many legal pad pages and spiral bound notebook pages and bound journals I have filled up throughout the years, read only by me… maybe some of those closest to me… the pages have to be numbered into the tens of thousands.

And that is not even close to being an exaggeration.

I wrote all of it down because it is my job. It is what I do. Who I am.

I type things up on the internet, post it, and hope for the best. Something built up in me on that plane ride back to Montana. A simple question:

Why had I not started writing a blog?

After landing in Missoula, after my little family was unpacked and tucked in for the evening, after many drinks and other things I dare not mention, after building up some courage, I typed into a search engine how to start a blog.

Now, over 200 blogs later, I can say I am a little bit of a success People are reading my work, at least.

Do I Still Have It?

All men make mistakes, but a good man yields when he knows his course is wrong and then repairs the error. The only sin is pride.

-Sophocles

Having my son, Vincent, in North Carolina for a weeklong visit is nothing to sneeze at. It has been years since I have spent this much time with him, and frankly, I find the whole thing daunting and miraculous. After years of street life, alcoholism, drug addiction, despondency, and a general lack of care about whether I lived or died, caring for someone - especially a child - should naturally feel strange. I even had to ask myself if I remembered HOW to be a parent.

Throughout my time wandering, I, of course, thought about my son. I also thought it was better I was not a part of his world. Too much chaos was included in my presence. It (chaos) was like a sidekick, both internally as a cruel commentator and externally, like a thunderstorm shattering silences.

If anything, I had a little luck on my side. My wife, Tara, and I drove the long drive up I-77 North all the way to Ohio to pick up the boy. I was told he had problems in the past riding for long stretches at a time. Two hours seemed to be his breaking point, and he would start acting up in his seat, kicking the seat in front of him and trying to unlatch his safety belt. These are just a few of the struggles I was informed of. Our 6-hour and 42-minute drive might have been "interesting" and trying. However, Vincent kept a smile on his face for most of the trip, laughing and clapping whenever we would drive into the long tunnels that had been cut through mountains. Their supernatural, orange-yellow glow of lights the length of the tunnel might have seemed magical to him... or spooky-fun in the Halloween sense.

He was very, very good during the drive. It shocked the hell out of me, for I have this terrible quality of imagining worse-case scenarios whenever I am told something might be difficult. So, I thought of screaming, scolding, and crying all through the several states we would be traveling through.

However, no.

He did not fall asleep, cry, or be a nuisance in any way. The three of us laughed and clapped and talked... well, as much talking as Vincent does, which is little to none. My son is eleven years old. He barely says complete words - forget about complete sentences - he makes a lot of noise, though. Mostly, his noises are drawn-out moans, starting low, then evolving, growing into a high squeal-like sharpness. These noises can be prolonged, lasting an hour... sometimes longer.

It can get beneath your skin. Test your patience. However, there is nothing for it. Those noises he makes, I believe, are part of his coping mechanisms - a rhythmic, guttural comfort he can orchestrate.

Vincent is autistic.

Because of that, many things in his life - like his noises - differ from that of a non-autistic 11-year-old. His needs are various and specific. On the other hand, he is very much like me: wanting what he wants - knowing what he wants - and wanting it now. In addition, I cannot deny the lack of patience is undeniably familiar.

I have been told more than a few things regarding his moods and manners to watch out for. None of them are good. Yet, already on day five of his visit, Vincent has expressed himself in very healthy, "normal" eleven-year-boy ways. When he knows something on his plate is a food he does not like or is simply not interested in, he will not eat it. When he is not tired, he will not fall asleep. However, thank goodness for this, when it is time for him to be in his room at night, he stays in his room, usually watching YouTube videos of his favorite movies until it is time to come out. Which, regularly, is at a decent time in the morning.

He may spend an hour or two in his room after waking up. It is, in my opinion, an acclimation period... a personal ritual to be performed before starting his day. There is a security camera mounted on his wall, and the 5-inch, handheld tilt monitor is never far away from Tara and me. Most of the time, he is locked onto and mesmerized by his phone, watching it while jumping on his bed or swinging in a special chair screwed in and hanging from a stud in the ceiling.

(Now that should be a column all of itself: his damn phone. See, Vincent does not simply watch his favorite film from beginning to end. He will find a sentence from the picture, rewind it, and listen to it repeatedly until my ears feel like they are ready to bleed digital, electric-colored fluid, and my brain will be on the brink of imploding. The phone is his "safety Blanket" – (an escape from boredom, discomfort, or anything else that may be disturbing him.)

Some things had to change in the cabin before Vincent's arrival. All the firearms had to be unloaded, stashed away, and locked up. My library was dissembled. All the shelves and books are now placed helter-skelter throughout our home. There is no order whatsoever. Vincent's room was where my library was. Tara spent hours decorating and organizing it. A solar system theme was her aim. Glow-in-the-dark sticker stars and planets adorn the walls. His entire personal space is secured and enveloped by the imagery of outer space-colored lights and comfort toys.

It is a neat room. Tara should feel very proud of herself. Vincent, oh, man, Vincent really likes it.

In addition, I do not mind the changes as much as I thought I would. Hell, I do not even mind so much sitting at the dining room table for every meal for a change. We are finally utilizing the very large - very heavy - dining table. A practice Vincent does not seem to have been pushed onto him with regularity. (The first night we sat down to dinner together was the most difficult - frustrating because he wanted to keep standing, not using flatware, and his insistence on pacing around while eating food with his hands was aggravating. Nevertheless, we got through it.)

He has gotten much better, though. He has been with us for nine days. His improvement - not just with mealtimes but also with many things - has been a happy thing to witness.

My son being here is a good thing. Good for Tara and me, Vincent, and I hope it is good for his mother, little sister, and younger brother. Having to raise three children is daunting enough, I imagine. Having a Special Needs child can be a great blessing but also a great hardship for a family of five. With it being Tara and me - both of us working from home - it might be easier and better for Vincent to spend more time here, in North Carolina.

That is not a slur on his mother's no doubt difficult successes in raising him thus far. However, I cannot deny MY home being able to offer him some different options, not to mention more one time he so very much needs.

When you see an opportunity to contribute something good to another's life, take it. It could be an opportunity for many.

Sweating It Short

It could be the heat, possibly high humidity. Maybe it is from all the pills I ingest. Would genetics have anything to do with it?

A big problem I have is sweating so much. Literally!

It is an ungodly amount. So much so that I have to change my shirt two – sometimes three – times a day. I don't even want to know the number of times per day when I have to push my glasses back where they belong because they are sliding down the slick decline of my septum from those rivers of sweat and the numerous occasions leaning forward; my glasses need to be removed to wipe dripping droplets of perspiration from the lenses.

Simple frustrations can evolve into a psychotic break where, completely against normal behavior, ugliness shouts and shouts.

I sweat as though my skin is too toxic to sustain liquids.

In addition, I will shout,

And

Shout.

Millions and Millions of Soft White Needles

No matter how often or how little I try, they are there. All of them. Either way, they are out in the open or hiding under the couch, behind the television, in my toothbrush. Hell, they are everywhere. All the long and short, soft white needles.

They fall from the bodies of the dogs. Hundreds of those needles fall with every stroke and pat on their body.

What is most funny: after sweeping up the floor, a white, soft mountain is gathered… it is almost as if you could make a longhaired puppy with all the fallen needles there are.

When I wake up, they are in my mouth, on my pajamas – everywhere. If they had a voice, they would fill up the city of Rome.

A million little soft, white needles… how precious they are when they are gone.

Newspaper, Worms, and Memories

Twenty years ago, I read The Athens Messenger every day. That was during my first (and consequently, my only) trimester at Ohio University.

That was a Presidential Election year. The year our slow, child President, George W., was reelected into his second, embarrassing term. That was in 2004. In addition, it was the first year I ever voted... the first time I felt what I believed in was worth believing in... That I had a voice.

Reading the local and national news daily was as important to me as my classwork, the manuscript that would become my first novel-length work, and far more important to me than, say, human relationships. At twenty years old, I felt my first surge of patriotism. Sure. That great high and energy have wavered, been alienated - even smothered to the point of temporary suffocation; however, I persist in reading the news.

Today - almost twenty years later - I find myself back in Athens County, buying fishing bait at Baum Tru-Value Hardware. Tara, my son Vincent, and I were spending a short, 24-hour vacation together at a house on Green-Up Lane in Long Bottom, Ohio. A pond, with several small boats, a raft, and the water stocked with black crappie, red ear sunfish, largemouth bass, and bluegill were at our disposal.

Taking my boy fishing seemed a good idea. It's not a bad way to unwind from the drive from North Carolina. I had never taken Vincent fishing before, which seems sacrilegious in a very absurd American way.

What sort of father hasn't taken their son fishing by the time he is eleven?

By that age, the boy should already be sneaking off on his own, staying outside all day, and coming home late to go fishing.

(There are plenty of things I need to catch him up on. I was absent from him for far too long. In addition, fishing is one of those things, dammit).

Movie Magic

That's all it is: magic. Entire cities can blink into an ocean… some existential tide of images coming from a screen, conceived by computers… wounds painted on bodies, digitally detailed to enhance the destruction of Self and Others.

Entertainment has always been the same. It has always been about escape; it is also meant to see things – know things- that are not part of an everyday occurrence.

It is amazing. It is magic. A safety blanket to ward off the wounds that could possibly happen.

Yet, the "real" world is out there.

It does not need to be magic.

All of it is real, in a way.

Back and Forth

My standard morning – back when I was living in Cleveland, at the sober house – was not exciting. When I was twenty-five, walking back and forth on a Cleveland ghetto street at six-thirty in the morning was not what I envisioned my future mornings to be. However, after I got a few (that means 5) cups of coffee and half a pack of cigarettes into my system, the walk became necessary.

Not so much for me.

It was a solid time to spend with my friend, the Urban Philosopher Rickolas.

See, Rickolas is someone who needed to SCREAM! Not in any mean (or literal) sense. He just needed to vent every morning to continue through the day with a stable mind.

I understood it. Hell, part of me was no different from him. *I needed to scream*, as well.

Rickolas needed to speak when there was no one else around to judge what it was he had to say. And, well, me… well, the hell can I say – I just listened to him.

I was elected the official listener.

When you live in a sober house, there are ears everywhere. And that is not simply paranoia speaking; it's true. In the kitchen, in all the rooms, EVERYWHERE – someone is listening. When you go to the bathroom, you are wondering if there are any deceptive digital devices recording the embarrassing mutters we (well, at least I can admit to it) sometimes have during the Call of Nature… or, just when we want to be alone, and the bathroom is the only space you have to claim some privacy.

Walking up and down the side streets of the neighborhood was the only way to have a private conversation.

And this morning, I am thinking of Rickolas. He most certainly had something he needed to talk about without the rest of the house knowing.

(I don't even know how I can – or should – write about this. I know that writing can be a form of exploitation, and I noticed people have been *reading my online columns*. So, I was going to try and do the best I could to keep this candid yet have candor.)

Here's the deal:

An elderly individual both Rickolas and I were close to was having medical troubles. This person could not eat due to his problems swallowing… something about burs in his throat. This person has a feeding tube. Food (in the form of a liquid not remarkably different from the product Ensure) must be administered into his stomach through a tube. Not only does he have to endure this, but he also just had surgery on his frail and brittle wrist, preventing him from feeding himself.

He is asking others to feed him.

We lived in a sober house. Our freedoms were very limited. This was/is not a professional care facility for the elderly. None of us drunks and drug addicts should be held accountable for caring for someone in such a desperate state.

How could we even handle such a situation?

I know. I know. I know. The standard thing to do is inform the Executive Director of the house about what is going on. To a certain degree, she – the Executive Director – knows about the unfortunate situation. She knows he is frail and has cancer, which he is going to die from. Professional help (i.e., a nurse) is becoming increasingly needed.

(She does not know he is asking for help to be fed from a resident/patient of the house.)

In the end, I know the situation is not going to end well. That is just the ugly truth. The man does not want to leave the place he has called "home" for the past however many years… it is a standard, tragic story. A story many Americans know. It is a standard story. But it is still a difficult, frustrating story full of grief and tragedy worthy of a cheap novel. Or a great novel, depending on the author.

Rickolas and I kept walking. After his "outburst" of emotion, after we walked up the road yet another time, I'd come to realize the houses lining the road were not that bad looking. One or two are empty; the windows are boarded up. But most of the homes have nice flowers and plants growing in pots. Shrubs are trimmed. The driveways are paved. I have been in much, much worse places over the years.

I'm sure the elderly man has been to some bad places, as well. Hell, I know he has. I wish I could tell him he can go someplace better. Better than the West Side of Cleveland we are in.

But it is not my place to do so.

Nor is it my place (or Rickolas's) to care for him. Care for him in the capacity required.

The Great Fear

He had to think up a story. The page was placed in the typewriter. That terrible white page – that blank page in all its torturous glee – looking at him, taunting him.

He knew he could write something, but just writing something to fill in a blank page is a jerk-off type move… writing something good – to have someone lost in a story is something completely different.

He knew he could type up a list. He could write a description of the floral-patterned wallpaper in the format of a poem… but he did not want to do that. He wanted a story. And the story is what eluded him. Something to start a novel, a short story. A novella – anything, as long as it had a beginning, middle, and end. Hell, even an idea to start a script would be something to consider.

None of those things were calling to him, though. He sat and typed every day about that day. There were some good entries and some good advice pieces, but there was that old Capote quote running through his mind: "That's not writing. That's typing."

He knew he was a writer. Sitting down at the typewriter with the devotion of one eternally damned and gifted – even when nothing came out – looking at the page, scratching down ideas on the many notepads and notebooks next to his Royal typing machine, he knew. He put the time in every day. He has even been published! More than once! He felt he was a legitimate literary professional.

There had just not been a story in a long time—nothing he could build upon.

What kind of writer went about his day without a story bubbling in his heart… in his mind? A story that needs so desperately to be written down, the fingers jolt from the sentences formulating in the mind!

He looked over at his spiral-bound steno pad, looked at the words jotted down in black ink from earlier in the morning:

A reform for reciprocity for the Looking Glass Personality.

What did that even mean?

He did not know.

More importantly, how and where could he use those words in his day's work? And he was going to work. Come Hell or Heliotropium Heaven, he would get one page written.

The blank page stared back.

There was the story about the serial killer. Stories about his drinking problem – his drinking problem, not the serial killer. The serial killer was sober.

The page kept smiling back at him…

There was no need to write anything grand. No fancy language. No moral. No profound insight into the Heart of Man.

He just wanted a story. He just wanted some pages that felt wonderful to write. Some pages would cool off for twenty-four hours, be reread the next day, and have the story continue.

No need to worry about typos. This would be a first draft. Typos were expected. Typos were like unwelcomed, surprising blemishes on a great lover.

And he was the King of Typos!

Rewriting something was a bit scary. That second draft business really brought the work into play. The editing… the expanding.

The second draft was the puberty before publication… or, more accurately, the submission process.

Maybe today was not the day to start *The Story*.

Today was meant for this. What you just read.

Holiday atop a Smith Corona

Memorial Day. Holidays had taken a difficult spin for me. For two years, in many addresses and many different cities, I was homeless. Or in institutions, jails, rehabs, and treatment centers, or hospitals.

Well, you get the picture.

The holidays just passed on by like any other day, without any kind of celebration.

No gifts. No feasts. No laughter from the young or smirks from the old. All I can remember from one Christmas was the sad and deeply scarred and heavily drugged-up faces… faces of those wishing they were somewhere else – anywhere else – and with someone else.

I remember thinking I was just grateful I was not freezing in the snow. Freezing fingers turning blue and numb, unable to undo my belt to take a leak… bad, bitter times that were hard with the cold.

This day, though, even though I was not with family or little kids to remind me of the *real* importance of life, I was safe.

At that sober house, there was a BBQ burning some burgers, hotdogs, and other delicious holiday meats. There were even steaks on the grill, smoking and sizzling away. Baked potatoes wrapped in foil, waiting for the butter, salt, and pepper. There were guys from the house outside, playing cornhole. A game that seems to be played in every rehab and detox center and treatment center I have ever been to. Some good things were out there, happening outside of my room, outside of the house. The weather is nice, bright, and cool. A perfect warm day.

There was no blood.

No dirt or grime underneath my nails or smeared across my face. No stitches or casts or the rash-burns of discomfort on my body.

It was just a nice day.

Was there some discomfort in my head?

Yes!

But even though I had some wandering, not-so-pleasant thoughts in my noggin, *I was safe*.

What will it be like in a year? I remember thinking.

If I were to be honest, and I had to be honest at that typewriter, I feared the future holidays. I feared them because I felt it would take a couple of holidays – a couple of years – to acclimate myself to my "Family" again. It would take time for me to be welcomed back… and for some, welcoming me back is not an option. To them, I am too far gone, too fucked up… I fucked up too much for them ever to accept me.

Not that I blame those assholes. And that is what they are: ASSHOLES! Hellbent on keeping a grudge. One of them has even gone so far as to steal a .410 shotgun my uncle left for me. This person lies and lies and lies it was left for him.

It wasn't.

My uncle left the shotgun to me years and years ago, before the little shit had a driver's license.

(I doubt the little pissant even has a cleaning kit to care for the antique single-shot.)

But that is how they are. They think they are always right and everyone else is wrong. They will lie, change facts from the past, and do so with the kind of self-righteous vigor of the Gestapo.

I knew it was going to take time for others in the family to accept me .. that is what I thought. To my grateful surprise, it did not go as difficult as I thought it would go. In fact, it was very easy. And all the love and protection I felt from them was still there… it was like it never left.

I very much wanted things to be balanced. But I have not been blind to the complexities of other's emotions, thoughts, conceptions, and beliefs. It is hard to "keep it simple" when a person is complex.

People, in general, are simple. People (in large numbers) are not difficult to fathom. One-on-one, on the other hand, is something completely different.

But it is going to happen. I *will* have holidays with my family again.

The bonfires with us adults sitting around, talking adult things which make us laugh at how ridiculous we are…

Watching people unwrap Christmas gifts.

Roasted turkey.

Colorful eggs to be hunted for.

IT HAS TO HAPPEN!

Hell, for the sake of my psychological well-being, it must happen. And that may seem selfish, but so what? Most of us desire a family to be close to on holidays and vacations.

It is far better than what was going on before on the streets.

Yes, I did the Long Kiss Goodnight to family… or so I thought.

That's right.

There was a time, while I was down and out in Youngstown, I thought I would never see family again. Part of that assumption was correct: I lost an uncle during my "bad" times. An uncle I loved very much.

He is long gone now.

Hopefully, my uncle is up in heaven, keeping the ice cubes cold for my stiff drink past the pearly gates.

I said goodbye to them, though, in my head.

That was a terrible thing to do. And I lived in a terrible way. It gave me even more of an excuse to mourn – to drown and to suffer and to drink and to use drugs.

Any excuse would work.

Now, even with some feelings of resignation, it has to happen. To see them. To be a part of familial normalcy. And to let them know they were always there… they were always with me, in my heart, in my mind.

I fear something great when seeing them.

Rejection.

Judgement.

They have the right to do both… but I know only two of them will.

It's a very sad thing. And they are sad men. But what can you do but pray for them to see the errors in their ways, which, I'm pretty sure, will never happen.

Their pride will not allow it. And pride is the ultimate, blinding sin.

It Can Happen...

There are very few times, I believe, when a person can officially declare they have everything they desired... back when they were a kid, that is.

Today - of all the dreary, rainy Mondays of my life - I am able to say I have acquired all (and more) that I so fantasized about as a child.

How did this happen?

Dreams and wishes can become a reality. HOW I managed to achieve mine is beyond my comprehension. I do not deserve the gifts I have, nor the few and rare people I choose to have in my life, and the home that I live in. It wasn't even five years ago, a rainy and dismal day like today; I was walking up one street in Mahoning County, Ohio, to turn and walk back up the same street. My clothes soaked through from the cold rain. My body shivered violently to warm itself... my feet wrinkled and sore in soggy socks...

God, those were desperate and bad times. There was no hope for me, let alone a future. It was all downhill and drudgery. Looking up into a grey, lifeless afternoon sky over Ohio, not wishing or wanting anything, asking myself:

WHAT IS NEXT?

WHERE CAN I GO?

I was not looking for a definitive answer. In addition, there was a much more serious question of what it was that I could possibly want. At the time, I wanted EVERYTHING. Everything I thought a good majority of Americans had. Things like a home, a car, a bank account, superficial depression - an existence away from the wet streets.

A little while after, while I was living in a sober home in Cleveland, I really, really started thinking about things, things, THINGS that I wanted as a child. Things to aim my energy at to obtain. Goals, if you will, to satisfy long-ago wants. Stupid things, like a Jeep, and I mean a REAL Jeep. Shaped like a box, tire on the tailgate, 4x4 JEEP! Ever since I could remember, a Jeep has been what I have always envisioned myself driving. Driving through deep, wet, and terrible mud puddles and ditches, driving up the side of mountains... to a child's mind - a very young, adventurous child's mind - a Jeep was the total ticket to vehicular freedom.

Item #2: A Winchester Model 1894, 30-30 lever action rifle. The reason behind this... well... I saw it in several movies where the good people use a Winchester 30-30 to stop the monsters. A childish fancy, but I was a child, and that is how I thought.

There is the weapon that will save the day and the vehicle to drive me into the bright and bleeding sunset into glory. Cue the music. Start the Credits.

Anyway, those things - those treasures - mean something to me. They are proof that not all was lost. To the contrary, much has been gained. In addition, it had nothing to do with a work ethic, a con, or a flip of the switch in the mind... it had everything to do with faith and a little determination. My relationships, my thoughts on life - EVERYTHING - have changed so drastically from those wet days in Ohio to where I am sitting now, writing this.

There was a little hint of fate, a whiff of luck, and the belief other people had in me that has made me the man I am today.

It worked - is working - for me. Not all has to be lost to find your place in this life. However, if you feel or know you are going down, focus on something small you so desire. It may save your life trying to obtain it.

A Day of Films and Words... Just Another Day

Just another day. I put a movie into the DVD player, turn on the television, turn on the writing machines, and listen to dialogue as the wonderful drink warms my brain to thoughts... thoughts on love, on life... of a guarantee – when there is no guarantee, other than the fact we are going to die – but, to be honest, the only thing immortal is the dialogue, the words.

Words.

Words.

Words.

Entire civilizations and societies have been built up by words.

Maybe one day we will be better at using them appropriately... maybe not.

Before the Drive

It is just another night. Another night of cleaning and putting things away before sitting down at the keyboard. - Work, like many things, manifests its own warm-ups and rituals. Tonight... tonight, it was watching Tara fold the clean clothes. Me, on occasion, folding along with her and putting them away.

(I have written about my pre-writing rituals enough; I will not go into any more details.)

Tonight, there was something different. I was placing some of the folded clothes in a separate pile. A large, green canvas bag was waiting open for me on the kitchen table. I was picking out what pants and shirt to place into the black mouth of the bag.

I was also thinking about a shower... what clothes to put on after the shower. What will look respectful? What would remain cool during this very hot summer season? It is the beginning of July, for Christ's sake! In addition, I live in the South!

You see, I am not going to sleep tonight. Alternatively, if I do, it will not be the peaceful sleep of the weary finally won. No. No. It would be a restless, fighting sleep. Probably full of uncomfortable thoughts, a breeding ground for unsettling images. The type of images I will not remember upon waking.

Before ten tomorrow morning, Tara, Winnie-Winnie-Winnie, the Ditch Dog, and I will be traveling up I 77 North. On our way back to Ohio. A short trip. Just an overnight at a home with a pond. We are picking Vincent up in the historical city of Marietta. The city on the Ohio River. The oldest damn city in the state. 1788, I think, was when it was populated and established. It is an "Antique-type" town - dubbed the "Riverboat Town - and I have driven past and through it more than once. More than half a dozen times, but I only stayed there once.

It was not a good family trip. It was during the beginning of a long and destructive fall for me. This was years and years ago before the South adopted me... before my Southern, gentlemanly way gripped and set root to grow what I am today.

Wanting to go back to that town - that town where Mr. Bacus himself, the pioneering individual who took it upon himself to bring some form of Law and Order into the town, could not have stopped me from linking arm-and-arm with the original Roman Bacchus, and the two of us, for some animal-reasoning needed to walk stupidly and drunkenly down the old streets. It was a bad trip. Not one I wish to reenact.

However, I will be returning to pick up my son. Afterward, we will drive out to the country to a house Tara found online.

Taking Vincent fishing is one of the plans I look forward to the most. There are boats - paddleboats and a kayak - included with our stay. It has been years since I have been on the water - in any way.

If I drown, at least I will be going out doing something fun.

As I am writing this, a dog - one of the Great Pyrenees - named Gretta is by my desk, looking at me... the expression conveys a longing and some questions: Where are you going? When will you be back?

"It's alright, girl. We are not bringing him back this time. However, we will bring him back. Do you miss him?"

This dog keeps up a long stare focused on me. Whenever I look away from my typing screen, there she is. Not in an intimidating way. Not this girl. Her presence was more meaningful... more extraordinary... almost spiritual in her way of arguing at the core of me. All with a stare. She seems to be picking and plucking away at things that have been bothering me for about a month now.

I have no answers for her.

Not yet.

Things will grow... or die on the vine. The point being: something has been planted in me, within my wife... even the darn dog - all ten of them, for that fact - have something growing in them.

Something is either going to come or be postponed.

We wait up here on the mountain for it, though.

The New Littles

“Now, you cannot get mad at me."

Immediately, my guard went up. When the first thing your wife says to you as you are just waking up is for me not to be mad... well, how else should I react?

"What did you buy?"

She said she had not bought anything.

Yet.

That is what I thought - YET.

Soon, she went into a "sales pitch" personality. Telling me they are so cute. They are a good, reasonable price. It really did not matter what I did or did not say; I knew when my wife's mind was made up. Her mind was made up. Her sales pitch was just a stepping stone - a preliminary pas de deux - before her inevitable purchase. In addition, she really has every right to buy whatever she wants.

But GOATS!

That is right. We have gone from collecting dogs, hatching and raising several different chicken flavors, and picking out exotic-looking fish for our large freshwater aquarium to GOATS! Nigerian Dwarf goats, to be exact. Little and furry pieces of fluff. DNA has to have some popcorn threads and strands in it because they cannot seem to help themselves. They hop, hop, and hop from one spot to the next.

Yes, she bought them.

Two of them.

We named them Hemi (Hemingway) and Gelie (Gelhorn).

Their first night, they slept in the damn second-floor bathroom where, in the bathtub, resides eight chickie-peeps. Silkies. Maltese chickens. That is what they look like when they are grown: like a Maltese puppy with a beak and a twisted and puffy, punk-like hairdo.

Later, I would spend most of an afternoon getting our garage/shop ready for both chicken peeps and goats. Tools needing put away, the floor swept up from the debris and treasures the dogs feel compelled to bring in from the yards, and setting up an area in the back half of the building - an addition-space just right for the friendly livestock we are accumulating.

"We can start up our own little petting zoo," I said.

"I know," said Tara, "isn't it great."

Great.

Right.

I held back a laugh but smiled at my girl, knowing she was happy. Sometimes happiness is as simple as acknowledging those you love to feel complete... even if it takes goats and fishes and chickens and dogs in order to accomplish it.

A Hum of a Conversation

"See you again, I'm sure."

He had no idea who (or what) he was talking to. No one else was in his smoky, pleasant, and warm-smelling room.

Nor was there a mirror on a wall or the back of the door.

Something had to be in there with him. He could feel a pink, soft, and pleasant presence near him.

It is always near him, letting itself known in times of sleep or when he is supposed to be sleeping. A good friend. One which will be there at his end… one that was there when he was being pushed to be born.

Ms. Winnie-Winnie-Winnie Goes to the Big City

She had to set a good example. She is "in training" to be a Service Animal. Her black and tan face, with those big, gold, and intelligent eyes, seemed to welcome the extra attention her mama and I were giving her. Moreover, she does not seem to mind the getup: a SWAT-team-like canvas vest, thick collar, and heavy-duty leash attached to her collar.

She is a big girl. Almost 70 lbs. Not even ten months old.

A big girl that fills up the back end of our Volkswagen Tiguan. In addition, how she rides in the vehicle is a blessing. Mostly, she stays sprawled out on the bench seat. Other times, she has her head held up high, sniffing at the wind coming through the cracked windows, wagging her tail at whatever mystery she is detecting with her whiffs.

Tara and I returned to Ashville, NC, to get some paperwork done, have a good time, and surprise/celebrate my Mother-in-Law's birthday.

Winnie-Winnie-Winnie is now our Service Dog. That was how we justified taking her along on the trip. Yes. Yes. We had to get her an outfit, and some papers needed to be signed - all of it worth every second of every inconvenience to make it happen.

She was a highlight of the trip. A good girl, she was. When Tara and I ate at a restaurant, Winnie-Winnie-Winnie was right there with us, usually under the table, with her nose sticking up in the air, smelling the great aromas and various smells of a restaurant.

Tara bought special treats for her. When she was well-behaved, Winnie-Winnie-Winnie got a soft, delectable, bone-shaped treat. When she was about to be not so good, Winnie-Winnie-Winne got some more treats. It was a trip of treats.

The only problem we encountered was the first night. The three of us were in a cafe, meeting a couple to have the paperwork signed, and some other dogs were in the building. Winnie-Winnie-Winnie's back hair raised two inches when she saw them. She would growl and bark her warning at them.

Winnie-Winnie-Winnie is VERY protective of her mama.

Maybe she is a little protective of me, as well.

What I do know is that she is ahead of the curve. However, after that first night, Winnie-Winnie-Winnie got over her over-protective postures and noises.

She remained a good girl.

I woke up in the hotel bed, my wonderful wife to the right of me. Winnie-Winnie-Winnie had squeezed herself between the two of us.

It was a perfect way to wake up.

No, I do not live a perfect life... mornings like that, though, are as close to heaven as I will probably ever know.

Late to Bed, Early to Rise

It was going to be great. Waking up a little after 5:00 AM was a great sign. Great! My head was full of joy. This is right. JOY!

This was my chance to start a day off well, ready for action, work, and accomplishments gained. Not the usual trudge, grudge, and groaning from getting out of bed. There was no need to down copious amounts of coffee, nor would I need to smoke half a pack of cigarettes in order to jump-start the heart and head into functioning order.

I woke up ready.

All I needed was some time to sit on the couch, collect my thoughts, and build up the bravery to sit down at the typewriter and begin the day of work.

After the introductory smoke was smoked and my drink refilled, I walked downstairs to my office, sat down at my desk, and looked at a blank page…

In addition, I stared.

And stared…

And stared.

My mind, my courage to create, ran away from me and hid in a dust bunny underneath a bookshelf.

Maybe I need another cigarette and drink that second cup of coffee, I thought.

Should I get the broom from the closet? Those dust bunnies and larger dust rabbits need to go!

I looked at the blank page, lit that second smoke, and, of all things, typed up a sentence: IF WE CANNOT LAUGH AT OURSELVES, THEN WE TAKE THE RIGHT OF LAUGHING AT OTHERS AWAY.

Habits and Addictions

Most of the time - almost all of the time - I have specific "needs," which help with my writing process. Well, heck, just to sit down to type, some of these habits and addictions must be implemented and fulfilled. Turning on the laptop or sitting in front of the mocking, black mouth of a typewriter, the spat-out white tongue of the blank page wiggling and wagging possibilities and ideas is easier to face with some creature comforts. For many, that strong cup of steaming coffee is necessary. Hemingway kept a lucky rabbit's foot and horse chestnut in his right pocket; many of us cleaned before sitting down to work.

I am guilty of that. Cleaning, that is. It is almost as if I have a fear of having to work on something I know will be hard, confusing, or, worse, unattainable.

"At least the office is clean and organized," I tell myself, feeling as though I accomplished something. Which, of course, I have... but not what was intended.

When the production begins, a drink must always be to my right. A perpetual drink, one that does not empty throughout the process of pecking away at paragraphs. An ashtray, also to my right, with a burning cigarette, is essential; the cigarette clamped between my teeth as I type. I drag the smoke in to think and exhale as my fingers hit the keys. I do not always hit the tray when I ash. At my best times, after the pages are complete, I will stand up from my desk and, like grey snow falling from the sky of a smoky metropolis, ashes fall from my shirt and lap.

Movies are also a big deal. Always (and I do mean always) a movie has to be playing as background, as a soundtrack to my motion picture of paragraphs. Most recently, I have had a 1988 science fiction horror film, Watchers, playing in the background. It stars Corey Haim and a great Golden Retriever actor. It was one of my favorite films growing up. It is a story about a boy and his dog - being hunted by a genetically created killing mammal. What's not to love? It is scary and cheesy, with a bad soundtrack and obviously fake blood, but I love the dog. A super-smart Golden Retriever capable of communicating with humans through barking (one bark for YES, two barks for NO) and is a very intuitive animal on human emotions.

This movie has been my go-to on films needing to be played while I write. In fact, I edited my entire third book to that movie.

All of us have our quirks. My need to write comes with quirks. I accept that.

Wow

North Carolina is bright from the late July sun. It is a hot afternoon- hot enough that there was steam rising from the mild, mid-afternoon shower the area was gifted with, and prayers of thanks were given to Willis Haviland Carrier (1902-1950), the inventor of modern air conditioning.

I was going to mail out some packages. Our local automotive repair and parts store has a shipping section in the shop to mail things out. To my dismay, I could not send the boxes and padded envelopes – all of them containing my latest published book – so I decided to drive across the street to a shopping complex.

Normally, I do not care to shop. Too many people. Speakers mounted on the walls and installed within the ceiling are too loud, playing music I do not care for; walls of bright, flatscreen televisions and hundreds of other electrical devices I have not the faintest clue how to operate intimidate me, leaving me with a feeling of inadequacy. However, on this particularly pleasant North Carolina afternoon, I wanted to buy something. In addition, going home to purchase something online would feel like a cop-out.

Dollar Tree seemed to call out to me. Yes, I thought a Dollar Tree might have something fun for my son, Vincent. Moreover, going broke in a Dollar Tree would be beyond embarrassing; it would be shameful.

Just before the toy aisle, at an end display, was a six-foot tall assault of various glowsticks, plastic glow-in-the-dark rings, and figures, all of them in many neon colors. Neon green, neon orange, neon red, and blue. After selecting three different psychedelic novelties, my shopping basket and I walked down an aisle containing a small Books for Sale section.

The selection was what you would imagine. Stacks of cheap Western novels, large, plastic ring-bound cookbooks, and, as always, the trashy Romance novels, with covers featuring an illustrated, muscle-etched grotesque wrapping his one arm around a woman looking as though she is fighting back an orgasm while her long, long hair is perpetually caught in some gust of fantasy wind.

In no way did I anticipate finding anything worth purchasing. Even a dollar has dignity, not to mention my time is valued. However, the spine of one of the books made my eyebrows rise. A white spine and black cursive scroll, written as if by a crayon, informing all lucky enough walking down the third aisle of the Wilkesboro Dollar Tree that Lawrence Ferlinghetti wrote a book titled Little Boy, A Novel.

No Way. No way in Hell. Lawrence Ferlinghetti! Co-founder of City Lights Bookstores & Publishers in San Francisco! Author of A Coney Island of the Mind, one of the bestselling books of poetry in American history!

"What are you doing here?" I asked for the book, taking it from the shelf. Opening the book to the Copyright information, I discovered the book was published in 2019, and the copy I was holding in my hand was a First Edition.

This is odd, my mind determined, curious if this was perhaps some genius, very elaborate prank. I collect First Edition books. I collect first-edition books published by Lawrence Ferlinghetti and all the other Beat Writers of the 1950s and 60s, all the way to the present. (I just, in fact, procured a Signed First Edition book of poetry by Ron Whitehead, Lifetime Beat Poet Laureate.

How?

Wow!

And why?

Finding this book is a sign, I reasoned… though I am unsure of what. That is a good sign, for sure. Maybe something in the universe wants me to brush up on my Ferlinghetti? Not that such a request is unreasonable or unwanted.

Needing to read more poetry and discover a little bit more about the history of the literary Beats seems like a hot ticket. First, however, I need to delve into the little novel from the Dollar Tree.

On the Night in Question

There were many, many paper mountains on his desk. Files are spilling out white what-ifs from the folds.

He looked at the screen, knowing a message must be written. For over two hours, this writer – this fate-placed Appalachian asshole – had been reading and reading the hundreds of thousands of words from poems and stories and journal passages.

There are too many mountains to climb in one night.

He was the adventurer, forgetting the map back in last life's cabin.

Did Hemingway Talk to His Typewriter?

Not only was it frustrating, but it was also being wounded bad enough to bark out a reproach – an insult.

"Fuck you!"

No response.

"I said, 'Fuck you!' and you are saying anything. Not even a clever retort."

Still, no response.

"I at least had something to say to you. I still have words. You have nothing unless I have something. Where are your words? Inside me! That's where they are."

A blank, blank stare.

"I'm sorry."

Still… the still, still, still stare.

"Tomorrow will be better. You will see. I promise."

Evaluation with a Turkey, Swiss, and Mustard Sandwich

What is it like?

Living on the mountain, what is it like?

"Every morning is not like a holiday or a moment to mourn. It is not a guessing game. I know where I am. I know the smells, the embrace of the bedroom, and the sounds. It is home… home, infinite."

You are going to stay and live the rest of your days in the cabin atop the mountain.

"That is my plan. Not only to live here but also to die here, in my own bed, with my own rules, cherishing all the memories – good, hard, blissful, and unique – till the moment the old, old bastard of finality comes to collect me."

Was life ever as good as life on the mountain?

"Never. Not even close."

You should finish that sandwich you are chomping on and get into that bed upstairs. Your wife is waiting for you.

"Okay."

Smuggling an Opossum into a Rodeo

After the shocking bookstore ordeal (don't worry, dear reader, there were some really, *really* good happenings and future plans set and brewing in Salem, Ohio's Bob's Bookstore) after dropping my son, Vincent, off at his mother's house in Ravenna, Tara and I ended up in a staring contest. The two of us were wondering if we should accept an invitation to join my cousin, Jenn, and her family to a rodeo taking place in Burbank, Ohio, or not.

It would be a thirty-five-minute drive from our location in Ravenna.

I texted Cousin Jenn, asking how much time was left for the festivities and performances. Assuring me the cowboys, barrel racers, deranged rodeo clowns, and bull riders would be going until well after dark, 10:00 to 11:00 at night, at least.

Tara barked at me:

"What's the Security like at the entrance?"

"How the hell would I know?"

And that is what my mind screamed at me! *How in the hell would I know?* Going to a rodeo was as alien to me as vegan entrees served in Venezuela. What could I possibly know about the security measures allotted to rent-a-cops and off-duty patrolmen for the safety of Ohio's wannabe cowboys and their families?

You might be wondering why my wife was so concerned about Security, ain't ya?

Well, she needed to smuggle Mildred, our refugee opossum, into the event. All throughout my years, years of sneaking booze, drugs, weapons, and explosive contraband into events onto airplanes - never had I imagined I'd be in a wildlife smuggling situation.

I agree. It was an exciting concept with unknown possible outcomes. The marsupial beast could escape! She could end up running amok, biting at the bucking broncs, the brutalized bulls, causing a stampede that would crush innocent spectators.

Headlines would read: "Broken Bones and Spilt Blood in Bullring... Oppression and Horror Caused by Opossum."

But certainly, we could not leave that animal in the Jeep. Enjoying ourselves in the grandstand while Mildred suffered darkness, loneliness, and uncertainty seemed ludicrous, not to mention out of my wife's character. She brought Mildred to the lunch we had earlier with my parents at a nice restaurant and bar called Coaches. Ms. Mildred slept in a canvas pouch/bag. My father, thank the Lord, was completely unaware of the situation.

Mother gave me the heads-up in the parking lot to keep Mildred's presence a secret because, as she put it, my father would "freak out."

That made me chuckle a little.

If there was any "security" at The Bukin Ohio Rodeo and Bull Ring, they were elsewhere - somewhere away from the entrance. Tara had her bag over her shoulder. Mildred did not stir. Not once did I see any indication any out-of-the-ordinary thing was bundled inside the bag - especially nothing *breathing*.

My little family and I joined a different part of my family that night. One I do not see enough.

We watched the competition, admired the animals, smiled, and laughed at the children pitting against one another in the ring.

Having to bring Vincent back to Ohio was not the most pleasant of things, not for me and especially not for Tara.

We needed the rodeo, Mildred, Jenn, her family, and everything else.

Everything else is always good medicine.

An Afternoon After Lunch

Hot and sunny afternoon outings, running errands, doing the doctor appointment catch-up thing for insurance.

His stomach was full of the Mexican Special plate he had for lunch. Tacos, rice, and fried beans weighed heavily in his belly, giving him that special feeling of being full without the weight-down of discomfort.

With a head still feeling light - his heart as heavy as his stomach - he thought back to the many Mexican lunches he had enjoyed... and not enjoyed. Lunches in Ohio, Montana, Texas, Alabama, Florida, and North Carolina.

All those states in the Nation... all those tortilla chips and spicy salsas and, in previous years, ice-cold *cervezas* (or *chelate*) to wash down the burn. It also, at times, washed away the struggles.

There were plenty of struggles he had to deal with. Back in those days, he *thought* he knew about struggle. Today, on a beautiful late-August afternoon, there was a *real* struggle. One that not only involved him. It involved many people. And an innocent life's future was in the brew bubbling in his brain.

Everything was wonderful. He thought of the last two years, years living in the South with her... an assemblage of times differing in joy, frustration, and advancement. There had been a sense of completion awarded in his new life. A peace, he thought, which could only be found in books or at the ends of sappy movies, just before the credits rolled and the room turned dark.

Now, that peace had been shaken. He did not know what was going to happen. Not that he ever did with any God-given situation in his life. But this one... this was one he not only wanted an answer to bury in the backyard of his brain, but he also needed one to move on.

It was the answer, whether it was in his favor or the complete antithesis of it, which would consist of one of two words.

YES Or NO.

A Day Years and Years Ago… Salvaged from a College Notebook

Nothing much to report today. Rain continues to flood the sidewalks and roads – micro rivers flow their lullabies over the asphalt and concrete.

Those young professionals hurry along, umbrellas over their heads, or papers and file folders transmogrified into umbrellas… the stomp, stomp, stomping into puddles, crossing the small creeks, still trying to chatter away some banal conversation about "the job." And we – as in me – continue to suck down and snort up my chosen poisons… not to mention swallow hundreds – if not thousands – of colorful miracles, always waiting the long wait for them to make their debut.

Maybe it's an encore?

At this point, what does it matter?

It's a gluttonous hope for oblivion. Not to mention heightened action…

They predict the rain is going to continue till the beginning of next week. Dismal are these days of Spring. Loneliness is following all of us home with damp realities, hoping to dry off into fantasies unaware of the storms.

"It's going to be a long day," I whisper to the glass.

Not that I get back an answer.

I do hope for one, though.

2:57 AM, In the Present

Somewhere under the dark stain of the building, a message was being dictated from somewhere far off.

The impulse to spray paint the words – the poetry – which had manically averted my thoughts from all other aspects of being was taking over.

Writing never seemed like the comprehensible – or responsible – decision for my future. However, whether it be homeless, soon to be painted – some poetry painted – on the side of a building, or in the comfort of a safe home lightly tapping away at laptop keys – *the words are always with me!* They always will be.

They need a home.

Just as much as I do.

Philosophy of the Game (More Notes from College Journals)

There are winners, and there are losers.

We all have been told that.

Some of the heavyweight pigs in the paralysis of bad luck laid their heads down on the bar. I imagine them wishing they had bet on the other team, playing a different game. Or wishing that they had never bet in the first place. Their loss will follow them all the way into next year… the year after that… and the year after that.

Spouses will continuously remind them of the edge, the mountain they aggressively walk upon. How soon the cards fall, revealing the addiction at its core.

Pains of the punishments and the lectures and the revelations follow the progression of defeat, judgment falling, falling, and falling down on them in spurts of insults and curses.

On the other hand, winners are the worst of the gambling junkies. By this, I mean they follow a surge of an indescribable, powerful illusion. One that is camouflaged and untraceable to an average person's mind. A thought that no bet is a loss, no price too high for their wallets, and where credit is an inexhaustible well that never has to be replenished.

It is when these bloodhounds lose their first really, *really* big mark - one they must pay. And it *truly* hurts their position. Then, as if on cue, the trend starts its progression: one loss turns into two losses, then three, then four, and so on and so forth into an abyss that prides itself on being the kissing cousin to hell. It is the memory of winning that keeps them going. It keeps the engine humming for the next game of chance.

The game of gambling has many distinctions, but none more brutal than the deceiving reminder of what it was like to win. It's not even the prize that is gratifying - just the cheers and self-knowing of beating the odds or following through with a gut feeling that turns out to be the correct play.

Justin was one of the many, weighing his head down on the bar, asking himself how he could have bet so high on the long, hard, and terrible shot that he did.

Brandon joined him in his misery. That should-be-solitary grieving process losers feel, and everyone knows.

There were a few walking out of the bar shouting and cheering themselves in glorious triumph. A collected celebration – a unity – of Them vs. Us, with Us bringing in the gold, the glory, the green.

The losers… well, the losers stayed sitting at the bar, feeling what coins were left in their pockets, wondering if there was enough for another drink. Some, most likely, were thinking about the wife they eventually have to go home to, having to explain where the paycheck went… where the rent went… why there are no armloads of groceries being carried in.

Another paycheck down a perpetual drain of losses.

"Hey, Smiley," Brandon shouted, "Will you buy me a round?"

At first, I thought of refusing the request. Then, thinking about my own dependencies, I questioned the harm in giving him some *spirit* on this atheist's night.

To commiserate with those in much need of commiserating is of no sin… none that I am aware of.

"Chris, give that man a shot. A double, if you will."

I raised my glass to the loser. A sign stating *good luck* in his now luckless moment.

I watched Chris poor from the bottle of Maker's Mark, hand it to Brandon, then mark it on my growing – never-to-end – tab.

After Brandon gulped the shot of whiskey in a shivering swallow, he stood from his stool and walked out with the rancor of an ashamed doctor.

Several moments passed.

Chris set a brimming glass of light-red drink, wrestling with ice cubes in front of me.

"On the house," he muttered.

I kept my eyes on the drink.

In the background, the sound of dirty, clinking glasses, plates, and silverware was being loaded into a washer.

"This is the end," I whispered.

A Long Time Ago Office

In my memory, the city always seemed to be at the mercy of late spring. Its weather is faded, cold, and colorful, filled with wetness that seeps into your skin, your hurtful bones. The wetness soaking the soul with sadness - a heavy wool coat of damp sadness and grey/green afternoons that form into a wet and shining night.

Many people find their way into a place where some comfort can be gained. But where do you go when those places close? When the doors are locked, and you are left out in the dark and nighttime lights over the streets?

There were poles, illuminated by heavy green neon, welcoming patrons into the Clove Pub. It was a form of light that both welcomed and warned. A light that comes from darkness… is borne of darkness, welcoming each patron with its warning.

I had been to the Clove many times. It was the type of establishment that, despite its torrid reputation of being a meat market, served good food in the late afternoon hours and maintained a downtrodden sense of unity. All the regulars know one another. We know what we drink, where we work, who our wives, mothers, and sisters are.

In a sense, it was a non-corrupt, political ideal of sousing-it-up with a low-class flare.

This is a dying city. Years ago, back when the rubber factories were still burning, families could be supported on the lowest wage/salary the companies had to offer. When the kids graduated from high school, many of them didn't waste their time with higher education. Their careers were all set and ready for them to clock in the Monday after graduation. However, since the fall of the tire plant, many – if not most – of those potential "lifers" on the line are heading the way of the jobs. Naturally, how can they not? Everyone needs to make a living.

Now, with the fleeing of the younger generation, the city council is trying to develop new ways to keep young, college-educated graduates interested in the city. Keeping the young money where it came from.

Dull, greying, and tired lawyers from the firm two blocks north sit in booths, scribbling away on yellow legal pads, long-haired blondes looking seriously into Art History textbooks, and the frat punks with pitchers of cheap, piss-bubbling beer… this is the cast of an afternoon in the Clove Pub.

And I am here to write about it. Every day with my notebooks and pens, sometimes a mini-tape recorder.

My high is with the passing from one page to the next. There are, when the moment is right, a good observation – a little hint of gold shifted through the banality. That is what I bank on.

Those lines go onto a page, maybe, or later, in the manuscript.

Shishosetsu

The "I-novel" to the Japanese is called *shishosetsu*. Although the writing I post is FAR from being a novel, they are autobiographical. Confessional Literature, if you will. And I try to keep everything grounded in truth. Exaggerations of how I react or when I confess a feeling are somewhat common. Everything in my stories is true.

Take, for instance, this story about Mr. M__ M__. Mr. M__ ran a tattoo parlor on the West side of the city of Missoula. I went in for a tattoo. Ironically, it was two bleeding M's on the inside of my right arm.

When I went into the shop - before I knew the man's name - I could tell something was off. He was sweating profusely, even though the air conditioning was blasting Northern Arctic spit-cicles - hence, VERY cold! So cold that I was shivering. It must have been 90 degrees that June day, and I was sweating from the walk from my apartment to his shop. My shirt was soaked through with sweat, and the cold air was not helping me in the situation. In fact, it was making me even more uncomfortable. If you know what I mean. I remember worrying I would not be able to hold still long enough for him to needle the ink into my skin without flaws.

Then I saw his hands.

And I understood.

They were trembling. Trembling in a way I was VERY familiar with. After a few moments of conversation (what kind of tattoo I wanted, what I was looking to spend), I had to say to him:

"You need a beer or something, man?"

He looked confused.

"It's alright, man, I get it. I drank more than a six-pack before I came in here. Now, do you need a beer? We can chill for a while till you feel better."

He looked *so relieved.*

I offered to buy a case for us. That is what came to pass—a case of Budweiser for two good guys looking for some release and laughs. We sat in the back. I sat on the special client's chair, sipping away, pleased with myself. Mr. M__ began to relax, sitting on his little swivel work stool.

After about three beers, he had calmed down. In fact, so calm and confident that he brought out a binder full of his original illustrations. They were done with pencils and, I imagine from the details, fine-tipped pens. What struck me about them was that every one of the images was an object of something "safe" or "comfortable." For example, one of the pictures was a couch - a couch with vines sporting deadly-looking thorns. Another one, the BEST one, was of a kitten licking its paw. However, the kitten's tail looked like a medieval weapon. Spikes grew out of the tail. And the tail was wrapped around the kitten's body.

I had to ask him what he started a drawing with.

"Oh," he said, "Which one?"

"Any of them. All of them," I replied.

"Well, I start off with something simple. They just kind of change the more I get into it."

Fair enough, I thought.

But I was sure a psychologist would have something to say about my new friend's distorting safe and cute subjects into things a nightmare would like to have in its inventory.

Anyway, we got comfortably drunk. He inked the two bleeding M's into my arm. He laughed:

"You know those are my initials?"

No. I did not know. But it made the day - not to mention the tattoo had more meaning. Its meaning - the tattoo's meaning, changed dramatically two weeks after I got it.

One day, walking past the shop, I saw it was closed. There was a sign on the door.

CLOSED DUE TO DEATH

What the hell?

I looked it up online. M__ M__ died in a car accident the previous night. He was drunk. Those are the only details I remember.

I do think about him and those nightmarish sketches. His initials are on my arm, for goodness sake. How could I not be reminded of him?

How can I not think about how he died? And all those nights, I could have checked out the same way.

I hope he is in a place where there are no more dangers and nightmares. I hope to see him again one day.

What it Means to Me

What does GONZO mean to me?

The definition, according to Merriam-Webster and the Free Dictionary:

gon·zo

ADJECTIVE

informal

NORTH AMERICAN **of or associated with journalistic writing of an exaggerated, subjective, and fictionalized style.**

1. bizarre or crazy: "the woman was either gonzo or stoned"

synonyms: **mad · insane · out of one's mind · deranged · demented · not in one's right mind · crazed · lunatic · non-compos mentis · unbalanced · unhinged**

2. Extreme, unconventional, or bizarre: ***gonzo artwork; a gonzo snowboarding style.***

3. Crazy, excited, or unrestrained: ***Fans went gonzo when the band came out.***

The definition - according to the dictionary - is simple, specific, and not hard to comprehend. It's not like, say, Nietzsche's essays on the German population's effects on Western culture.

However, for me, the word represents a complex philosophy of life. Not just "demented" or "unhinged". It is a form by which to live.

Controversial?

Sure.

It really is. However, as weird and unconventional as it is, I identify with it.

A form of surrealistic, accurate, blown-out-of-proportion perceptions on life and environment, politics, writing, and reportage... a little bit of fictional spice in a recipe of what is in front of you.

1971 was the year GONZO appeared in the dictionary. Of course, things have been GONZO for a lot longer than that. Mark Twain, Norman Mailer, and Truman Capote were all using the technique.

Hell, even Theodore Dreiser experimented with concepts of breeding fiction with reality in his 1925 novel, *An American Tragedy* - that famous, long work based on the murder of Grace Brown (1886-1906); a depressing story with the elements of human horrors that, not surprisingly, was turned into a literary masterpiece.

Years later, Truman Capote produced his most accomplished work, *In Cold Blood* (published January 17, 1966), on the 1959 murder of Holcomb, Kansas farmer Herbert Clutter and his family.

Capote called it the "Non-fiction Novel."

Norman Mailer continued this new and exciting avenue into literature with his mega 1979 Pulitzer Prize-winning true-crime work, *The Executioner's Song*. Once again, fiction mingles with fact. Gonzo and a Gonzo lifestyle throw all the extraordinary "possibilities" and "assumptions" in with the sometimes stranger, hardcore facts. On more than one occasion, I have been told stories about myself that are nothing more than exaggerations. For a long time, I perpetuated it, entertained it, and wrote about it… it was a controlled, creative chaos. Bred from ONE simple fact.

But that is not what Gonzo means to me. For me, Gonzo means family - a gathering of diverse people, some of them acting a little more outrageous than others; some of them not outrageous at all, just different, passionate, and connected.

All of it was fine-tuned and defined by one Dr. Hunter S. Thompson.

Because of Dr. Thompson and his powerful, diverse writing, activism, outlandish lifestyle, political insight, and contribution to controversy, I'm honored to have people in my life I never would have had the pleasure of communicating with. There is also the courage he inspired in me to not be perfect. He inspired me to use a steel trap snap grasp on what separated me from my calling: to be a writer. Ever since I read Hemingway, I wanted to write. To write professionally, that is (I was writing for “fun” long before the notion of considering it a vocation ever came into play). To get the music down with words!

To do it RIGHT! And true! And with abandon.

"Tis strange - but the truth is always strange; stranger than fiction, if it could be told..."

That Romantic, limey bastard George Gordon Byron wrote that in 1823. I wonder, though, did he ever consider what hybrid livestock would emerge by breeding the two?

Well, I guess that question has been answered. Some pretty good writing has been born from it. A lifestyle philosophy has been born from it. I am happy and proud to be a part of it, no matter how small my part may be, it's true to who and what I want to be, accomplish, and dedicate myself to.

Daily Reminder

This morning, I have family on my mind. I'm writing this on my laptop, sitting on the couch. Right in front of me, on the coffee table, are pictures framed of me and my wife and, of course, family. The one that I have been looking at the most is a 4 X 6 color image of me & two of my kids. Normally, I prefer black & white photos, but this one is special - full of color.

I'm in the middle one in the photo, my arms around my son & niece – two of my kids! It takes something pretty strong to beat having two smiling children on your lap. I have three beautiful children, and I love them. Two of them are from my sister and brother-in-law. I know. I have written down those lines before. But some words will always need to be said, written, blasted off into the heavens, and cherished.

I look at that picture, thinking how small they are. The children my son, no bigger than a big toddler, was probably around that age when the picture was taken. My niece - I have no idea of her age during the time the picture was taken - darker skin, a twin-like reflection of my sister and her husband... and, oh, how beautiful she is.

She just got an award for having a 4.0 grade-point average all year from school. (I NEVER got a 4.0. I was lucky I graduated.) She is also on the track team, plays volleyball, and survived several illnesses this year. She is truly one of my heroes.

Along the line of frozen moments on the coffee table are pictures of my wife, Tara, with one of her dogs that passed away this year. There is a photo of my uncle, my mother's brother, who passed away several years ago, and two images of my son - all of these photos remind me I have a past.

I have a unique family. All of them contribute to making ME unique in my quirky way.

Family is a funny thing. At this moment, an aunt of mine is close to shaking hands with the Good Lord himself. The doctors gave her a month left to live. I received this information from two other family members I have not seen in years. Their hearts were open enough to make contact with me.

That means so much to me!

I know all I can do is keep all of them in my prayers, which I do. And to keep communicating.

I'm a sucker, one with a big heart. Some days, I wish I wasn't... but I can't help being what I am. Anytime anybody casts a line out my way, I go for the bait. Wanting to be in the lives of family and select, non-blood-connected people is a good thing.

(Balancing that with my hermit-like NEED for isolation has caused me trouble… both in the past and present.)

My thoughts go to all families going through the hard times of life. I wish all of the good people around me could have the blessed life that I have had. A life I am still living.

We all need to pay a debt. We all owe a life. That is the price of having a life... but we don't need to cash that check alone.

Another Move

I will start this column with an apology.

I AM SORRY!

To all my daily readers, I apologize for not posting in the past few days. Somehow, in the course of the chaos of getting our home and property sold, my laptop was damaged. I'm writing this first draft piece on a legal pad. And there are two more waiting to be typed up once my laptop is fixed or I buy a new one. I don't even know what the hell is wrong with it. Tara is getting ready in the bathroom. We are going to Best Buy to see if the damn thing can be reworked, reprogrammed... or whatever it is they can do for it.

Some damn strange things have been happening in my world - well, me and my wife's world. We hope to be moving again. That is a strange thing to hope for. Moving is a pain in the ass.

However, we hope to be moving to Our Final Home! That's right! When/if this property sells, that will be it! It will be the home I die in.

I look forward to my hermit years, writing in a log cabin, enjoying life and cooking, making love to my wife, landscaping, having family and loved ones visit... being happy... I look forward to helping others when they need or want a place to stay...

True, my life has been tattooed in controversy - that was something I wanted. Now that I am older (not old enough to be wise, but old enough to recognize wisdom), TRYING - screw that, DOING - a kind thing is better than shock. Sure, shocking people *may* cause them to think, but I would rather help somebody instead.

When you think about it, hurting someone or shocking someone is easy. It's much better to help someone. The smallest of things can make a difference. Even if it's something as ordinary as making dinner for them (hell, just OFFERING to make dinner for them) or listening to what it is they have on their mind. - THAT IS HARD! Yet, it is easy to do if you try. And give yourself a reminder of the hard times in your past - times that might have been made easier, or even saved, if they were fought and figured out with a friend, family member, or even a stranger.

Lighting a fuse, openly doing drugs and drinking, fighting, carrying a shoulder bag full of "questionable" items - all those things were easy for me. Now, to continue being the person I have grown to become is a vow and dedication, and some days, it is hard.

But, dear Reader, it is in so many of us to be kind and good. You just have to ask yourself what type of person you need - or want - to be.

In The Past

There are things in my past I am not proud of. All of us do. Some of those things are very difficult to talk about, let alone remunerate the impact they have had upon me and the people in my life.

What I REALLY want now is to be a force of something positive whether it is through writing, an action, generosity - whatever! As long as it is good, that is all that matters.

A reader sent me a message the other day, a picture on Facebook. The picture was of a piece of paper, and written on that page were the words, "One day you will tell your story about how you overcame what you went through. And it'll be someone else's survival guide."

The Reader, dear Ms. Andrea, wrote to me, saying she read that and thought of me.

It was more than touching to read those words. It was a validation of everything I'm trying to do.

No, I am far away from stable and "ok" or even properly healed, but I am better than I have been in a long time. Yes, every day is a struggle to NOT go back to the hell I knew as home for so long. Today, I was given the archetypal choice of following a friend down the path I know all too well or trying to help him... help him through honesty - and, like all honesty, it had to be pushed through my personal conceptions, my personal desires - which broke through what the RIGHT thing to do vs. what I would LIKE to do is all about.

I will not lie: there are times I would like to "party" with abandoned impunity - where there would be no consequences for my actions. Incidentally, that is not the way my world - or anyone's world, for that matter - works. There are health risks, personal loss, and psychological complications ranging from depression to mania to psychosis.

Don't get me wrong, I am still no choir boy. I love doing drugs & drinking... that's the problem! And it became such a problem, my body & mind NEEDED all of it. Needed it more than people, books, goals, a life... everything else was in a grey & drowned background.

So, I got help. Help when I REALLY wanted it. That is the only way it COULD work: when I WANTED it.

Am I perfect and happy all day, every day?

Of course not!

Some days are very hard. My wife and I have our struggles. There are times I think about that bottle of vodka, that first line hit of cocaine... my goodness, everyone deals with stress in their own way. Everyone celebrates happiness in their own way, overcomes pain in different ways, and soothes the soul in solace in different ways.

I try not to hold onto the "character" I was for so long.

It came time to Man the Hell up.

We shouldn't always be looking in the rearview mirrors. There is too much in front of us to keep our eyes on.

In the Age of Anti

"A state of war only serves as an excuse for domestic tyranny."

-Aleksandr Solzhenitsyn

Now, a little about Mr. Solzhenitsyn: he was, above all other things, a writer.

"Aleksandr Isayevich Solzhenitsyn (11 December 1918 – 3 August 2008) was a Russian novelist. One of the most famous Soviet dissidents, Solzhenitsyn, was an outspoken critic of communism and helped to raise global awareness of political repression in the Soviet Union (USSR), in particular the Gulag system."

That's a chopped-up version of what Wikipedia.org has to say about him, anyway.

And why am I bringing this up?

It has to do with that quote… and a conversation I had with my wife earlier today. We were talking about the many things our generation has gone through. – What we have experienced. Everything as stupid as Y2K to the burning and crumbling of the Twin Towers. The COVID-19 pandemic, Colorado's Columbine High School massacre, and countless other school shootings… an Opiate Epidemic that has claimed more lives than several wars combined.

No!

Suffering and hardship are not competition.

All of us deal with it in our own way, in our own time, when it is time. History writers and their books, I predict, will have some interesting things to say/convey about the Age of Anti. And that is the title I have given this moment, the people that have lived during it since I was 22 years old: The Age of Anti.

I'm only 38 years old. That is 11 years more than what I wanted. Yet, the daily miracles still happen. Sometimes, it takes me a moment to sit down and think about it… but they are there. Today and tonight had some rough spots to them. Every day has its hard times. Just like every day has its wonderful moments.

My wonderful moment today was waking up. Waking up next to the girl I love.

What was your wonderful moment, dear Reader?

Grueling Grudges… Not Worth It

"As a writer, you should not judge, you should understand."
Ernest Hemingway

How very true that is, Mr. Hemingway. On the other hand, a man can remain angry. Just because you don't judge doesn't mean you can't be angry and hold a grudge.

Recently, I had to let go of some of my anger, swallow my pride, and be the "bigger" person. I had to let bygones be bygones. It was not an easy task, but I was forced to consider whether my pride and anger worth the unhappiness of my wife.

Well, it sure as hell was not. So, I had to bury the hatchet, sorta speak and put the ill feeling I had towards someone in a hole.

Now, I know I am being vague, dear Reader, but the details of this occurrence are irrelevant. The point, the moral, of the whole damn thing is I came out the better person. It's NOT forgiveness. No, no. It's not letting the person that most comforts and understands me suffer over crap that has happened time and time again and will happen more and more… I will not exacerbate her stress just to prove a point.

It's better to let it be. What is the point of holding a grudge?

That is a serious question I want all of you to ask yourselves. Take it from me: life is too short to stay angry, to continually remind yourself of being wrong that it corrodes even the slightest bit of serenity. And if you happened to be on the smelly end of the stick of a situation – when you need to vent, write about it, buy a punching bag – ANYTHING!

Don't let the BAD THINGS fester & grow & rot your day.

You never know; that could be the best day of your life… and you can fuck it up by being blinded by something trivial.

6:08 PM

It's coming to that time again, dear Reader - time for me to buckle down again and put everything together. I must edit the hell out of it and submit it: another book manuscript.

Part of me is NOT looking forward to it. It's embarrassing, sometimes, to sit and read what you have written, fixing the god-awful grammatical errors and misspellings. But, what the hell… it's part of the job.

It has been a very busy year. Getting married, moving several times, adjusting to MANY changes, putting my first book out, and, the hardest thing, beginning to be a part of other people's lives again. I never thought I would be in a position where I would receive so many messages and calls. Going from low-down-nobody-broke-and-homeless drunk to a semi-decent citizen with a wife, home, and a profession seems sublime. For goodness sake, I have identification, dogs, a truck, a bank account, credit cards… hell, I even have a WALLET. And clothes! Boots that don't look as if a pack of starved hyenas has had at them! Things to do and people to do them with.

Tonight, for example, my wife and I are invited to spend some time with another couple. I'm being SOCIAL. What the hell happened to me? I'm put together enough to be out in public without shame.

It's all very good. It's what I have wanted. Sure: things must be on my terms… I'm malformed that way, but it is happening. The dream is coming true!

Why is it coming true?

I'm not sure. I sure as hell don't deserve it. I'm sure as hell not going to turn my back on it.

That feeling you get when you accomplish something you are passionate about… there is no other feeling like that. When I first printed out my first manuscript/novel-length stack of pages, I was soaring. You sure bet I was soaring and full of pride. I have done a lot of drugs and drank hundreds of gallons of booze and beer, but there is no high better than the one you feel holding your completed creation. – Something you did and constructed from nothing. A story! A story made beautiful with your mind, time, and persistence.

This year, I got to hold my first book in my hands.

What a killer-rush-bitch of a sensation!

I keep going on and on. Every day, something needs to be written. Every day, something has happened worthy of time spent on the keyboard.

Remember: If it's important to you, always be faithful to pursuing it, whatever it may be.

One Last Walk Through

I have written about it before. But there is something different – something has drastically changed – about the movie theatre I went to as a kid… as a teenager… as an adult.

Cinemark Theatre in the Carnation Mall in Alliance, OH – a haven for me and the mecca for so many wonderful childhood memories - might be closing down. Hell, the whole mall might be pulverized into the ground and rebuilt into an obscene monstrosity of a store that would kick-start my social anxiety into high gear, if ever I attempted to go in, that is.

The theatre might be going with it. Just another mound of grey and colored heap of rubble, just another location where a large ghost embraces memories of the many.

My wife and I were visiting Ohio two weekends ago. We stayed at the same hotel we always stay at, which is attached to the doomed and haunted Carnation Mall.

(And, yes, that is how I view it: DOOMED and HAUNTED – a ghost-of-a-shell of what it used to be, filled with memories and memories and memories. I can almost SEE them. I can almost see the little me walking into the quaint and favored KB Toys, excited I had a little money in my pocket from pulling weeds or cutting grass.)

I had to go walk around the mall and try to see some of those ghosts I'm so fond of. Those dark walkways, gated-up storefronts – stores I remember from long ago: Walden Books, K-B Toy store, Pocketchange… all of them are gone now. And the very building in which they were housed will be gone.

I have become very accustomed to saying "Goodbye" to things and people from the past. I thought certain things would always remain. Carnation Mall and that little movie theatre… were places I thought would remain.

My mother took me there in the summers before I could drive. My older brother took me there before I could drive. I went to the theatre and saw *Jason Goes To Hell*, the last two *Lethal Weapon* movies. My brother and I had this thing when it was summer to see the stupid *American Pie* sequels at Cinemark… a whole damn childhood - a lifetime - is going to be reduced to rubble.

Not all things about growing older are fun. As a matter of fact, MANY things about growing older are no fun.

I am happy I got that one last walk through the mall. I'm glad and grateful for all the wonderful memories I have.

Having to acknowledge and accept all things naturally have an end is wise. But that does not mean you have to like it.

Two Places to Remember

There have been more than a couple of columns highlighting the good memories from my childhood. Here is another one. I'm sure many Ohio-born children can relate to and remember the hot afternoons and hot and humid nights of this particular State's summers. We also remember some of those days and muggy nights spent at two places called SeaWorld and Geauga Lake.

The first, a labyrinth of mass aquariums, which created habits for many creatures of the sea to be viewed, both in natural habitats, exhibits, and even live performances by Shamu, a large orca that would shoot himself out of his deep, performance aquarium, snagging a fish from a trainer's hands. Mamu was also an orca part of the act; however, I don't know much about her. - Shamu was the star, character, and household name performer. In fact, several orcas have played Shamu. The original Shamu died in 1971.

This was SeaWorld. The theme park/marine zoological park - still with several of the parks in operation in Orlando, San Diego, and San Antonio – the intriguing and magic-like experiences, for me, was in Aurora, Ohio.

(I think I was first there on a school field trip. My most memorable time there, though, was when my mother took me and my two siblings there as a surprise one summer. Spending the night in a *hotel*, even.)

For me, SeaWorld was a place to be merry and enjoy being little, in a place full of huge underwater creatures. Maybe I was eleven when I went on that last visit to the park with my family. It was the park across the river I had my eyes set on: Geauga Lake!

That was the theme park dedicated to the teenager in me. Going all around the park without adult supervision, roller coasters to get the adrenalin pumping, causing mischief. Yes, yes! From summer vacation after my eighth-grade year, up till I was 16, I went to Geauga Lake several times a summer. The first time, though, that I can remember, my Grandpa Locke took me. Unfortunately, I don't have many memories of that day. I just remember he was there with me, and we rode the Raging Wolf Bobs, an old, wooden roller coaster erected in 1988, permanently closed in 2007. This led to the deconstruction and sale of the famous ride. It was put up for auction and sold for $2,500 on June 17, 2008. For six years, the dismantling - as you can imagine - was slow; the completion of the project was in 2014.

There were many other rides at Geauga Lake that grasped that nostalgic, warm, and yet distant part of my brain - that part of the brain where things are never REALLY forgotten... it just takes a little jump-start of a reminder, and then - BOOM! You can remember smells and colors and days never before thought you would – or could - remember. Or, more accurately, ever NEEDED to remember.

While writing on this subject, just a moment ago, I remembered a television commercial that regularly aired one summer. It was for Geauga Lake's newest attraction: the Texas Twister. A commercial depicting a poor farmer, in a shocked-like trance, driving his tractor in the field and plowing a large concentric circle, symbolizing the ride.

It was the first ride of its kind in North America. And it was one wonderful s.o.b., spinning large spins in the air, rolling over and over, back & forth, swaying to the demands of its hydraulic gods.

The Big Dipper, the Double Loop: all these were big-name, must ride roller coasters at Geauga Lake. It is all gone now. The park is a ghost-like piece of property at the corner of the city. It looks like some shooting location for *The Walking Dead.*

It's sad. It was such a landmark for Northeast Ohio.

I'm not sure if there are any plans to restore it.

If it can be restored, that is. It's just the brittle skeleton and cracked concrete reminder of what once was. Fortunately, I have the memories... maybe that is how some things should remain: just a memory.

Dusk, Last Night

Yep. I did it. I did what I said I was going to do. The magnum rounds looked good going into the pond, and the fireworks looked good exploding in the sky above the pond. It was my way of saying goodbye. No! There is no closure. Not yet, anyway. But last night's send-off for my friend had nothing to do with closure. More like my expression to what was bubbling inside me since reading that awful text, he was gone.

Tara was with me. She was by my side, knowing I was hurting inside. I hurt for the wonderful woman and child left behind. I hurt over knowing it's going to be a long time - and in a different life - when I see my friend again. There was also the FEAR of hurt. A hurt I cannot – nor want - even to imagine. That would be the hurt of losing my wife... or if she lost me.

I know we are not built to last. I've known that for a long time. I know some of my lifestyle choices have shed some years off my life, but I made a vow. A vow that I would never leave my girl. With that being said, I'm not going to leave this life.

That is the plan, anyway.

It is one hell of a vow, I know, but one I would like to stick to. And why not? She has treated me better than any partner or lover ever has. If all she wants for me is to stick around, I think I can, as well as SHOULD, accommodate her.

This is going to be a short write-up of a long feeling of loss. Not many people from my past have stuck around long enough for us to care for one another.

My friend had that effect on me. He was the type I imagined would stick around and grow into the role of Old Buddy of Mine.

He was good and damaged and funny and irreplaceable.

The skies had explosions in them. The night roared with the power of my .357 for him. In memory of him.

Chiweenie

It's something that just happened. Well, maybe not. Maybe he planned it all along. Conscious of taking his time, working, and worming his way into my heart. At first, I wanted NOTHING to do with it. Didn't want a thing to do with his smelly breath, ugly little body, the tiny and sharp nails growing out of his paws, or his clinging, annoying presence.

I even gave him a title:

DOG THAT DOES NOT BELONG TO US!

That's right. For weeks - maybe months - I did not know this dog's name. Didn't want to know. He was not *our* dog. It was my mother-in-law's dog. That is what I was told.

(We just fed him; he spent the nights in our home. Tara pays for his care, his vet bills, his food... but he is not ours'.)

He's a Chihuahua/Weenie dog mix: a Chiweenie.

One night, back in December, I thought: *enough is enough*.

I put him outside, hoping he would go home. My mother-in-law has a doggie door. Surely, he would go back home. It was cold and wet enough outside. Surely, he would venture back to where he belonged, back to a warm, familiar environment. This hope lasted two days and two nights.

It ended the morning Tara was leaving to go to work. She told me the little brown pain-in-the-butt came out from underneath some chairs we had on the porch.

That is right, dear Reader, he had been sleeping on the porch, waiting for one of us to let him in.

For some reason, I couldn't take it. Tara says it has something to do with me being a good man. Now, I don't know about all that. A dog is a dog. Nonetheless, he did not deserve to be cold, hungry... or unloved, even though NOBODY on this wonderful farm likes him. He is not that different from a large, brown, and needy rat. A rat with bulging, panic-stricken eyes.

Nowadays, every night, he is at my side, on the couch, belly full, and he is taken care of. He has even graduated to sleeping in the bed, completely covered, usually next to me.

There is a saying that every dog has his day... well, this dog, now, has his life. And his name is Rider. Sometimes, when I'm calling for him, it's "Rider of the Storm" that's echoing throughout North Carolina Darkness.

"Just a Couple of Things"

When I suggested we go out for lunch, I did not mean to work well past lunch. By the time I was ready to take a break from writing to eat something, it was well past five. Not that my wife minded. She has been very supportive and understanding about my work being important to me.

It was raining and grey and cold outside. The opposite of what one would associate with Southern weather. It makes it easier to stay inside, to continue working when it's dismal out.

We were on our way to the restaurant when she pulled into the parking lot of the grocery store.

I was confused, which is normal, but I do not remember her saying we needed anything from the store.

"What are we doing here," I asked.

She smiled.

"Oh, I just need to pick up a few things for that pasta thing I want to make."

A few things. Remember that as I continue this tale of woe.

Let me share something with you, Reader: we were in that store for an hour. The whole time, without shame, this little lady kept telling me we were not grocery shopping.

We only had an entire cart overflowing with everything from three large frozen pizzas, bags of veggies, disinfectant spray, coffee & creamer - you name it! We were GROCERY SHOPPING. By the end of it, we were laughing at the absurdity of all our purchases in that little cart.

My wife could not even get to her purse at the check-out line to get her discount card out. The purse was being crushed by various items of varying weight & importance.

Oh, and I forgot to mention she likes to buy in bulk. So, even if it were just "a couple" of things to pick up, that small cart would still have been filled with enough food to feed the lower East Side of Cleveland.

And I told her I was going to write about it - our little comedic adventure, buying just a few things.

I swear, other shoppers were probably questioning our sanity... or sobriety. People shouldn't be laughing that much about getting groceries. And we were shamelessly loud.

We still ended up going to dinner. I had some of the best teriyaki chicken wings of my life. A Saxophone playing Latino blew the blues out of everyone in the place with fast, loud, and funny music.

After that, we came home.

She went to bed.

(Having to wake up at 3:30 in the morning requires some sleep.)

Me...

Hell, I stayed awake.

There were some things I needed to write.

New Additions

There are many things I did not expect to happen in my life. For example, I never expected to live in the South; I never thought I would have that wonderful day when I held my first published book in my hand. I sure as hell never pictured myself restraining a mother goat while my wife clamped suction cups and hoses to her, milking her for all she was worth for colostrum/milk.

That is right, dear Reader, we had a goat give birth yesterday. One male, one female. It's terrible: we gave them names, bottle-fed them, and they spent the night in our home.

(They were not nursing on their mama; it was cold last night... do the math. They stayed inside with us.)

It was an odd thing: two of the dogs took to the little, cute baby goats. Our old Lab, Hersh-Hersh, acted more like a concerned Mama than the actual Mama, trying to get to them, licking and whining over them. Bucky, the beagle, also acted as a furry bodyguard of sorts, watching over the babies as they slept on a large, padded dog bed.

It was an amazing thing to see. All the animals, all the species, including my wife, fawned over the little babies... knowing they were babies. And I'm no exception. Yes, yes. I will admit the little guys pulled at my heartstrings.

It just could not be helped.

They are just too damn cute! As well as hopeless. They needed some help. I'm a little guilty of being protective. I might as well admit it was MY idea to bring them into the house last night.

And that is the end of the story. Something short, true & sweet.

May Mahem & Laughs... A Birthday... Crashing A Wedding... Fireworks

Who would have thought my Wife and I could have so much fun? Well, I KNOW we can have a lot of fun, but yesterday – WOW! It was really amazing. And that's what yesterday was – FUN and AMAZING and GOOD!!! It was Tara's XX birthday (that's right, I'm smart enough not to reveal a lady's age), and I think I was more excited than her to celebrate.

(For years, there was NOTHING to celebrate... nothing that mattered to me, anyway. Having a bottle – many bottles – was my celebration, how I celebrated, and how I declined.)

When I woke up, it was still dark. Tara was sleeping her sleep of peace, and the night was cool, damp, and shining from the moisture of a night's rain. As usual, after getting myself together, I turned on the television, snatched up a legal pad and pen, and started writing... writing a birthday letter. That is different from a love letter, but not by much.

(I'm not sure what that difference is... it's just different.)

After the letter was written, printed, and sealed up in an envelope, there were her presents to wrap. Maybe that is why I chose to write first. I'm no good at wrapping. I'm sure there are some people out there who would argue I'm not good at writing… but writing is a more common, daily performance for me. Hence, I was more comfortable doing that. However, I was bouncing from one thought to the next on WHAT to DO next – what SHOULD I do next?

I was excited.

It was my wife's birthday. It had to be good for her. She is so good to me; she deserves the good birthday I hope she has.

Anyway, after the crude and ugly job I did at wrapping, she awoke. I was so excited to give the gifts to her that she had to slow me down.

"Honey, I need COFFEE first," she said.

(Totally understandable.)

After the coffee, after about an hour of adjusting from sleep to waking, she opened her gifts. Of course, after the morning merriment, we had to go hunting.

(My wife's odd obsession with finding a baby opossum has grown to an epic endurance contest. One in which she is determined not to fall behind or fail. What's a fella to do about that? My advice: support your wife when it comes to rescuing anything that is classified as an infant.)

Shadow Line Vineyard

With no success in finding any opossums, we started questioning where to go for lunch. At first, Tara wanted to go to a brewery she liked.

(It was, after all, HER birthday! Whatever her heart desired, I was game.)

On our way to the brewery, on back country roads, we saw a sign for the local vineyard. There were several colorful balloons tied to the sign, swaying and bouncing off one another from the soft breeze.

"Why don't we stop here?" she asked.

Like I said, I was game to whatever she wanted to do. The balloons, however, made me a little uneasy - it was too perfect. With it being her birthday, balloons on a sign... was IT a SIGN we should stop in? Or was there some private event going on?

Shadow Line Vineyard is a family-owned operation. It began in 2012, when the vineyard was first planted. It's located at 2550 Shadow Line LN, in the very rural area of Granite Falls, North Carolina. A place/location/state I now proudly call HOME. 2015: things were well on their way to becoming what it is today. February 13th, 2016: the doors were open.

I didn't know any of this going into the beautiful building leading to the tasting room. I didn't know what I was expecting... I do know what I saw when I sat down at the bar was not what I imagined.

Imagine for me, if you will, a curly and green-haired female, geriatric Joker, amidst middle-aged men in business suits. That's what was surrounding me. And all those bastards were wearing neon green neckties!

It was foolish, but I asked the bartender if it was St. Patrick's Day.

NO!

We had just walked into a wedding party! Not only was there a professional photographer there, along with the wine and catering, but *both* my girl and I were in jeans and tees and boots. (Tara, to my great pleasure, was wearing a Ladies' red Gonzo V-neck shirt.) I was sporting dirty, big pig farmer's boots, my hair was wild, and sunglasses covered my surprised and big eyes.

This is going to be interesting, I thought.

Best of all, one of the things I admire in that wonderful girl who claims me as a husband is that she walked proudly, fearlessly, and happily toward the newly betrothed and introduced herself. Not only was the couple kindhearted, but they were also gracious and generous: they invited us to help ourselves with the food.

We sat outside. It was a beautiful and warm day. Tara loaded a plate up with two types of pasta and some spicy buffalo chicken wings. She had ordered four sample glasses of wine.

(NOTE: Tara highly recommends the sweet white apple wine.)

It was a free lunch...

Why the hell not? Gifts come in many forms by many different types of people.

On their wedding day, the couple gave a wonderful birthday gift – KINDNESS TO A STRANGER!

It was more than that. There was a shared sense of celebration about the whole thing. Yes, I felt like a jackass because I looked like something that was dragged from a feeding pen; however, there was nobody there I saw batting an eye at my get-up.

Before we left, an invitation to return anytime was given. I imagine we will go back to visit. Supporting local businesses is important to Tara and me. The vineyard is about four miles away from our very front door. Shoot. I've walked farther than that for things less special than the time I had at that vineyard.

(I'm actually surprised it took us this long to go and give our support.)

Later On…

Dinner at her mother's place went well. The whole 24-hour period of her birthday went well. And it ended with a blast!

Literally.

As we were picking up some things for dinner, some random things that were required to smoke an entire chicken - apple-flavored, hickory woodchips, and such - I found a stand that had a small inventory of low-level fireworks. Smoke grenades, party poppers... nothing grand or great, like mortar rounds with a firing tube, rockets, or firecrackers. I did spot something that stuck out from the harmless novelties. It was a Boom Box!

Okay, for those of you who don't know anything about fireworks, this box is an aerial, 500-gram shot repeater. Twenty shots in the box. Like most fireworks, just light the fuse, step back, and enjoy the show.

That's what I – WE: Tara, her mother, and her mother's boyfriend Shane - did. We watched as an exodus-like carnal of colorful sparks and flames and explosions lit up the sky, shooting out in two-shot bursts from the Boom Box.

It was an appropriate ending to a happy day.

And the Beat Goes On

If they try, anyone can write about anything they know something about. Even if it is a deplorable, desperate, despicable, or even beautiful concept or story - people should write about what is inside of them, maybe even what they *wish* could be inside of them. Perhaps, maybe, write about what they are afraid to write about.

One of the wonderful things about America is *we have the right* to express ourselves. Whether or not people agree or disagree with it (also one of our rights) is irrelevant.

I'm not sure WHY I am writing about this...

Maybe it has to do with two conversations I had in the past week? One was from a wonderful and kind woman who inspired me and even paid me to read when I was in elementary school. The other was with my mother's cousin, though I call her – and identify her – as my aunt.

Naturally, each one had different reactions to my first book. My aunt has not finished reading the whole thing. She told my mother it was too "difficult" to read. - She has emotional reactions to the ugliness depicted in the pages. Terrible and hard things from my past are a major part of the text, almost a villain-character waiting to clutch onto you with the turning of a page. I can't say that I blame her for feeling the way she does. It would be hard for me to read about such hardships endured by someone I care about. On the other hand, there is kind of a happy ending (or, more accurately, endings) in the book. The sections go from dark to darker to better, leading to a good cross-section of growth and acceptance.

(Not that I'm trying to ruin anything for any of you that might read the book. It simply surprises me how different the reactions have been.)

This other lady, the grandmother of my good childhood friend, not only read the whole thing but wrote a book report about it, asking my family for my contact information to send it along to me.

How much I look forward to reading it.

Whether it be good, great, mean, or kind, to have someone take time out of their day to respond to what I do... well, it is a compliment. For years, I have written to people whose work I admire, and ALL of them have been generous with their time enough to respond to me and answer questions. It has been one of the great and good gifts in my life. Being acknowledged by people - hell, anyone that is a positive force - is a blessing.

A letter.

An email.

Even a phone call can change a person's day... for better or woefully worse. So always be aware of what you write. If you write to be hurtful - purposefully malicious... you don't have to send it. If it's purgation you seek, by all means, write it down. It does not mean it needs to be shared.

(Wow! I'm *really* starting to sound preachy.)

Memories and Lizards

Disappointment is an ugly thing. Expectations can be a fierce and daunting and cruel thing. (Never have expectations. In the long run – or at least statistically - you will be let down.) I'm not sure if I wish to dedicate this column to my amateur philosophies or to tell a story.

I'd much rather tell a story.

Back a long, way long time ago, before I knew anything about anything, my grandparents had a hunting lodge in Alabama. I might have been five... maybe six.

What I remember about the lodge was a bear rug on the wall. And there was a screened-in porch. I'd catch green lizards on the screen that darted up, sideways, and down and could disappear as fast as I could see them. My great Uncle Hud taught me how to fish for catfish. He liked to eat the eggs of the fish. I remember thinking I would never do that - eat fish eggs! What a disgusting concept. I'm 38 years old now. I have eaten fish eggs. Not the worst thing I have ever eaten. But most definitely not my favorite thing.

There are a few things that got me thinking about that Alabama hunting lodge. 1) We, Tara and I have a large, wooden parcel box. It sits about a quarter mile away from the main road, where our regular mailbox stands. With the weather getting warmer and the sun shining away throughout the day, reptiles are starting to come out. Snakes, lizards, and turtles are prevalent on the property. And for some reason, lizards like to hang out in and around the parcel box… probably the warmth of the wood, maybe the protection from predators, but they enjoy that parcel box.

Plestiodon Anthracinus. That's the little bugger's name. The Five-lined Skink. There are 13 different types of lizards in North Carolina. I'm pretty sure the ones hanging around the parcel box are the Five-lined Skink.

It always gives me a jolt, no matter how much I prepare myself for it. When I unlatch the door to the box and swing it open, half a dozen of the little things move with a snake-like slither very fast to safety. I mean, I know the little suckers are going to be in there. It's not like I am afraid of them. It's just unsettling having something move so fast over the deliveries. Just like I am not afraid of snakes. Being surprised by one, on the other hand, can make anyone jump a little.

Another thing reminding me of the lodge: yesterday, Tara and I visited a cabin. For some reason, the kitchen brought back memories of that Alabama lodge of my youth. That rustic and wild feeling of a cabin can be a good shock to stimulate the memory banks.

Things like smell and taste, even colors, can do that: make you remember the things forgotten.

Some things I don't want to remember; however, certain songs, certain smells - hell, traveling down certain roads can take me back in time.

I know the smell of hay will always remind me of the home I have made here, along with the smell of my wife's face cream.

Memory, like anything else, can be a good or a bad thing. Then again, everyone knows that... I hope.

Time for a Turtle

You never know what a morning can bring you. There is coffee, breakfast, check the news and the emails and Social Media posts - the normal things. Remember, though, I do not live a "normal" life.

I was awake all through the night. Nothing was wrong. It was just one of those nights. I moved from the bedroom to the couch, watched some television, then went back to bed. When I could no longer deny I was not going to sleep, I got up.

Tara was already awake and ready, lying next to me on the bed. That's right - READY!

It was morning. *Time to look for an opossum!* Get the special first aid kit, the box specifically, IF we find anything…

(Trust me, dear Reader, my wife is a relentless, motivated woman. When she wants something, she works for it and is dedicated to getting it. Hell, she waited and waited and waited for twenty+ years to wrangle me - and she won! So, the desire for her to save a baby opossum (or twelve) is going to happen... well, one way or another, she is going to save one or a dozen.)

It was time to get on the road. We needed to go to the store, anyway.

It was a beautiful morning. In the South, morning time is a time of color. Right now, everything is turning green and lush and also BIG, defying winter's grey and chilly past. The hills and mountains seem to have grown since winter - trees stretching out their arms with fresh, green leaves and the grass growing like mad in the fields.

Mornings are strange for me. I'm barely awake, barely human - it takes me about two hours of ingesting medicines and cigarettes before I am a "functioning" human being. Now, I am riding along the backcountry roads, looking for fresh roadkill. Still a stranger, WANTING TO FIND fresh roadkill.

I know, I know, it sounds awful and hillbilly, depraved and freakish. My wife is a humanitarian. A deeply, hardcore *I-don't-give-a-damn-about-what-people-think-humanitarian.*

She is going to save the baby opossums by God. She is a force on a mission, taking her husband along for the ride.

I have been on some weird trips before, but nothing compared to being half-awake looking at animal corpses on the side of the road, cars, and trucks with their drivers giving me the odd, bewildered expression that borders on confusion and disgust is kind of entertaining.

This is my life. I think to myself while prying the stiff-from-rigor-mortis legs of a run opossum corpse, only to find out it's male, saying words beneath my breath that would rot rose petals as effectively as anything dead on these roads.

One more let-down.

We start driving back home: our sanctuary/farm/haven.

We were driving, traveling up Highway 64, rounding the curves, and bullshitting about how odd it was there were so few animals killed on the road - almost complaining. Both of us were born and raised in the Ohio countryside. Where we grew up, it seemed there was a dead deer, raccoon, or dog every quarter of a mile on any number of back roads.

And that's when we spotted it!

It looked like a rock to me—a medium-sized rock in the middle of the road.

"TURTLE!" My wife screamed.

We turned the vehicle around, driving back to the "rock."

My wife got out. Traffic whizzing past her. She knelt down, picking up this thing that was, literally, on the yellow line painted on the road.

She was right. It was a turtle! And the poor bastard had been run over, the shell broken and chipped. He was alive, however, and responding to touch. I would take my finger and place it on the middle of his clawed, webbed feet, and his leg would move.

It was a Box Turtle. His head was safely protected within the shell. He was hurt badly, though. Tara and I had a conversation about me putting him out of his misery. This did not sit well with my wife. Desperate times call for irrational and desperate decisions: she called a vet. That's right, dear Reader, she called a vet. Completely shocked, I was told the vet would TAKE A

TURTLE IN! Yes, yes. We drove an hour away to a veterinarian's office for this Box Turtle to go into surgery.

Boxey.

That's the name my wife gave to this crushed creature.

We left him at Dr. Dunning's office in Morganton.

What will happen? Will he survive?

I don't know...

What I do know is I witnessed a wonderful woman who values the lives of animals. Hell, she values the lives of animals more than she does most people. And I don't blame her. I had never heard of a Box Turtle hurting anyone, never read about a chipmunk starting up a shooting gallery of gore in a public pavilion.

I will keep you updated on his progress if there is any. And if there is not, I'll inform you of that, as well.

A Sad Peaceful Ending

Well, dear Reader, the turtle, Boxy, did not make it. We received a text message from the vet. He stated there was too much internal and external damage to the animal that nothing could be done. So, in all the good that is decent in humanity, they put her down.

She did not have to suffer for weeks on end, which she could have if we had not taken her to the vet.

That's right, it was a female box turtle. And all I can feel good about the whole damned thing is - we *tried.* I guess that is the moral of the story. We went above and beyond to save her - a TURTLE, for crying out loud! In the end, though, some things cannot be saved. We must accept that. And that goes for so many things in life.

On a brighter note, I wish to thank you, Readers, for the messages you have sent me. Both currently and in the past. I even thank the haters. Yes, I have haters writing to me, as well. I try not to let it bother me. The only solace I have about this is the recognition people are reading my words. Whether they hate my words or love them is irrelevant. My job is to engage a reader. Scratch that last sentence. My job is to WRITE, stay sane, and feel the joy you feel when the word is put down…… with honesty and delight.

Ever since October of 2020, my dedication has grown. Keeping up with this online column/blog has been a lot of fun. Seeing people read it has been a wonderful and humbling gift - one that is new to me.

In all honesty, my favorite things I have written have been about the "normal" things of everyday life. Everything from grocery shopping with my wife to picking out a Christmas tree. And, of course, reminiscing with you Readers about some of the wonderful things in my childhood. Some of the terrible things from my past have been made better and easier to live with BECAUSE sharing and writing them down has made me understand some things a little better.

(No worries. I'm still an idiot. An idiot that needs a great deal of help.)

It has been a real blessing to confess and breathe with all of you out there. And I'll be first to admit that not everything I write is good, but I hope there is some good found in everything I write.

With that, I bid you *adieu*. I'll write to you all some more. All the days and months and years I have left, I'll write and share.

Good session.

A Much Better Year

Today was Mother's Day. I talked to my mother on the phone - twice. Everything is good between us, even though I was an ass for so many years. Last year, on Mother's Day, I got her flowers and a book written by John Fante titled *Full of Life*.

Both my parents came to see me. I was living in a sober house in Cleveland, OH. It was better than the previous Mother's Day. That is a night I do not want to write or talk about. All I will say is I did not talk to my mother that year.

Never in my life did I anticipate how wonderful life could be, let alone be able to share it with others.

Now, to change the subject, I'd like to talk about something MOST mothers do not like to talk about: Tattoos! That's right, the ink and art beneath the skin.

I got a message last night. Let's say her name is Ms. K__. Ms. K__ shared with me how much she misses writing to her great uncle, her grandfather's brother. She wanted to go to New York City with her grandfather. Unfortunately, he passed away. However, Ms. K__ has been a Yankees fan since she was born.

She has several "memorial" tattoos on her, reminding her of her loved ones who passed away.

(I have several of them, myself.)

Everyone needs reminders. The tattoo behind my right ear, **7:13** is dedicated to my son It was the time he was born. Why I was so insistent on getting that, I'll never know. But it seemed important to me. Behind the other ear is **V. R. R. L...** Those are my son's initials. I have **WOW** tattooed on my wrist, which I say is MOM upside down. And I have many more. Most of them are attributed to something literary, such as the City Lights Bookstore & Publishing logo. I got that tat in honor of the many authors they published. Authors would have gone unnoticed or forgotten if not for Lawrence Ferlinghetti and City Lights—writers such as Charles Baudelaire, Antonin Artaud, Allen Ginsberg, and many more.

As the song states: "Mutilation is the most sincere form of flattery."

Some may think tattoos are a form of mutilation. Some may think of it as a form of decoration. It does not matter which is right or wrong... if there is a right or wrong on the subject.

What matters is what is right for the individual. And what the design - the tattoo - means to them.

Do I regret some of the ink in my body?

Hell, yes.

But they are a reminder of my past. Something, at certain times, I do not wish to think about; however, I cannot – will not – deny that my past was something that happened. That my past is something I must deal with and learn from MUST be important to me.

ALL of us should remind ourselves that the past HAPPENED. No matter what we do, that devil-dog of a demon we sometimes call memory barks at the heart of us at the most improper times. Yet, with a tattoo, those times – both good and bad – are always there, making us LEARN from that one moment in time.

Thank you, Ms. K__, for making me think about these things, not to mention giving me something to write about.

An Email

Earlier this morning, I checked my email. I received a message from one of my mother's friends. She reminded me how, years ago, she gave me a pocket watch as a gift. A gift to remind me about time... to be aware of it, and to acknowledge its importance.

Well, after years where "time" was wasted, I now appreciate it for what it is. I also acknowledge how little time you, me, and everybody have in this world. Back when I was 19 (maybe 20), I wanted to be dead before the age of 30. I wanted to experience so much life in one quick and vast swoosh. I'd be gone before reaching the good Golden Years.

Well, right now, I cannot deny *these* are the good years.

Time!

What a strange thing it is. And it's strange we are the only animal that acknowledges it... that we are aware of it.

It's also amazing to me how much has happened in such a short amount of time. My world has completely changed during the past year. Everything from how I view life, relationships, devotion - even my writing has changed. All of which took TIME to digest and get accustomed to.

So many people - the society in general - are used to things NOW! I feel guilty in this category of impatience. When I want something, my patience is lacking. Not that I'm spoiled... well, maybe a little. Anyway, for the longest time, the thought of waiting for ANYTHING seemed stupid. I was going to be dead soon. Why wait for anything?

Things have changed, now... a little. It seems like everything is coming up so fast. All the changes and moves and growth have come as fast as fate-like bullets fired from fate's magnum.

Life is fast enough without impatience.

That's something I must remind myself of every day.

For All of You...

There are wars everywhere. Just read the papers, look at your iPhone, ask a stranger... war is everywhere. However, the most important one is the one you fight inside yourself.

That war in me, which I thought was gone, is a persistent pain in the ass. It's a damned annoying thing to fight with yourself. And to fight with things out of your control. Time is both a cruel and precious thing. There will be several personal matters I am going to share with you, dear Reader, if you don't mind.

I love writing. No matter how much I have threatened and tried to abandon it, I always seem to find the words working their way all through my head. There was even a point in my life when I hid it - hid pages I was working on, in secret, for reasons I'm still not clear on.

Why do I keep doing what I do?

It's who I am.

That is the jerk-off answer.

I need to be honest. I try to be better at what I do because I NEED TO BE SOMETHING!

For years, I just wrote and wrote. No audience. And none wanted. I just wanted to get it right. To put the fucking words down that were/are in my head onto the page... to get them down RIGHT - that was the goal.

Things have changed now, yet have stayed the same. In 1849, French writer **Jean-Baptiste Alphonse Karr** wrote *"plus ça change, plus c'est la même chose*"– the more things change, the more they stay the same…

Yes, yes, yes - I still write. I had to drill my way into the core of why. Why, now, do I continue? Not just because it is who I am and what I do. I'm talking about the core of why.

And I found an answer, and I'm going to share it with you.

It's for my family. I NEED them to know I'm not just a pill-popping, alcoholic, violent-tempered, crazy, and out-of-control failure in life.

I want to be good at SOMETHING! This is what I want to be good at.

I have good communication with my mother, my sister, my special aunts, and several cousins. And I want to thank them all for their support and belief in me. I'm very proud of all of them and give recognition to the same blood that pumps through our veins.

However, I must accept a very cruel and ugly fact: for some people, the past will always influence decisions. To decide to speak or not to speak to me. I have several members of my family who fall into that unfortunate category of "the past" controlling their decisions. - They are very proud and unmoving in their morals.

Who could argue with that kind of iron-made spine?

Well, I can. But I am mature enough not to.

Time & Time &

Time will keep on clicking in its horrible yet beautiful rhythmic beat. But there is only so much time. Believe me, it will end for me, for you, for everybody. Fear and Pride are nothing to hold onto. I have neither of those qualities. I have regret, elation, and a sense of belonging.

In my mind, I can see myself as a 70-year-old punching a mental heavy bag, still trying to achieve something from residual emotions from the past. To get it out! Hell, I see myself winning The Pulitzer.

I know I will never live to be 70.

I know I will never receive The Pulitzer Prize.

But I will keep fighting for BOTH.

Maybe that will be enough to get a letter, a phone call... a message.

All I know is I won't stop punching.

Getting Old

It is not uncommon for me to wake up in the middle of the night. However, it is uncommon to wake up in pain. My mind informed me to wake Tara up. That is what happened last night.

Now, I could not get comfortable. I tossed, turned, and rolled over... it was a major pain in the ass. More appropriately, a pain in the back. And that was the second pain of the night. All night long, I was fighting chest pains... I'm thinking of heartburn.

Tara was concerned. She is a complete professional, and trying to argue with her about medicine is as futile as trying to argue with my father about the digital read-out devices on an airplane. - It's just pointless. They are the best at what they do, and I have to listen and trust in them. I also accept when someone is deserving of credit for their accomplishments.

Let me tell you more about my morning:

After waking up, I accepted the terrible discomfort, and Tara gave me the once-over. She asked where and what kind of pain I was experiencing. She felt my back, I imagine, to feel for something - some muscles knotted and gnarled up.

"We are going to get you checked out in the morning," she said. A kind of managerial demeanor about her.

She was worried there was something wrong with my heart. To be honest, with the different lifestyles I led in the past, it would not have surprised me if there *was* something wrong with my heart.

But that was not the case.

In the morning, we were on our way to E. M. S. I thought we were on our way to see a doctor. No, no - *she* was going to be the one to check me out.

She entered the E. M. S. building like a little medical queen: walking in, telling the employees what she was going to do, not asking them. TELLING them! It was an amazing site.

My respect for her, which is high to begin with, multiplied even more. She walked into *her* world... and she was not going to take any shit from anyone, nor was anyone going to tell her no!

I had to ask her how this was happening... how was it possible for her to walk in, demand equipment, use it, and - now this part is interesting - no cash changed hands. No bribe or compensation to anyone.

"Those kids were my students," she said while laughing. "I'm the one who taught them how to do their job. None of them are going to say no to me."

Alright!

I can live with this, I thought.

All that was wrong with me was that some muscles in my back were injured, and I had heartburn.

The Golden Star for Best Performance of a Wife on a Wednesday Morning goes to Tera Locke.

Something Special, Something Unexpected

I sent some packages and a letter out yesterday. Just some gifts, and a letter to a magazine. I'm not sure why I feel compelled to GIVE so much away. Maybe it's the feeling of being blessed... wanting to share the abundance of happiness with others is not a bad thing, in my opinion. And getting mail is something I find very special... even when I know it is something I ordered for myself.

I remember a time when I feared getting the mail. Bad report cards from school.Or a bill I knew I could not pay. Hell, the worst, a summons. All of it seemed bad. Anything with a return address that I did not recognize, I threw it out. I was that paranoid! No joke. And if I did not throw them out, I hoarded them. A whole canvas bag of unopened letters from college loan debt collection agencies, credit cards... I can go on and on. I won't lie to you: I was terrible, irresponsible, and unreasonable.

For a long time, trying to find a point to such nonsense was beyond my reach.

Times are different now. Growing up took a long time. I can admit to being a slow learner... or just a stubborn pain in the rump.

What I can impart to anyone is the knowledge of discovering your problems may be a large, imagined construct of your own making rather than a manageable reality.

The only way to find out is to open those damn envelopes. Have some courage and faith in believing not all is lost… not everyone is out to get you.

Then there are the days you get that surprising, unexpected parcel in the mail. A gift from a loved one you have not seen in years... or a card saying "thinking of you."

It is worth it! All those days, weeks, months, maybe years of the same dismal drivel, then, suddenly, BAM! - a gift!

Those are the good days.

Those are the days we remember so clearly; maybe, if anything, it is for the brightness it brings to a secret darkness – a secret fear.

No more worries about the mail. Not anymore. There is more goodness in my life than any one piece of paper dimmish.

I think it was Charlie Chaplain who said something to the effect that nothing is permanent.

"Nothing is permanent in this wicked world. Not even our troubles."

That is what the good ol' hobo, Mr. Chaplin, had to say. To be honest, it's as good a philosophy as any. I think we should all remind ourselves that from time to time…

Listening to the Dishwasher

That is what I have been sitting here doing: listening to the damn dishwasher, wondering how it is that I got so lucky to have a dishwasher.

I have also been going through my files. I have enough material, new & old, to come out with *Notes from the Night Desk: Volume 2*. It's hard to believe... to have this amount of work, and knowing I had the dedication to do it, to stay on task to finish it gives me worlds of hope.

I have been reading a collection of Ernest Hemingway quotes and thoughts. *On Writing* is the name of the little book edited by Larry W. Phillips. In the book is this passage: "If a man is making a story up, it will be true in proportion to the amount of knowledge of life that he has and how conscientious he is; so that when he makes something up, it is at it would truly be."

Maybe that is why I do not try and write as much fiction anymore. I don't think Hemingway truly did. He wrote about what he knew - what he went through. Kerouac did the same thing, and Bukowski and Mailer did as well. Faulkner was right: Truth is better than any fiction. Or maybe the bastard was wrong. Maybe fiction is really the truth...? We just label it as fiction to avoid lawsuits.

The point being, for me, anyway, is I feel I should write about what I know. I'm not gifted like, say, Joyce Carol Oates, Stephen King, or any other fiction writer for that matter. Or, maybe I don't care about such things anymore. I have a small audience of readers who seem to find my words and life entertaining enough to keep reading. That means something to me. No, no, I may not come up with a great written idea as Jules Vern had done, but I keep up the work. I keep up the honesty.

Volume 2, dear Reader, will be a little disturbing. SECTION 2 of the book will delve into a darker part of who and what I was ten years ago. The writing, language, content, and style are going to make you keep on reading or throw the book down, damning me for the content. What the hell... I have to keep this wonderful thing up.

Sometimes, to keep going further and further on in life, you have to run in the other direction. Let's all keep that in mind as we keep going.

Waking Up to Thunder

A thundering and grumbling in the skies... I wish I could say that is what I woke up to. It wasn't, though. It was not the vibrating roar of the heavens that awoke me from a sound sleep or tears from the angels raining down on the world. - NO! The shivering, panting, and panic from several dogs woke me this grey and bleak morning. Two of those dogs - the rat terrier and the Chiweenie - were hiding beneath the covers, scared as a bag full of cats going into a hot tub. To the right of me, laying lengthwise, was the Chiweenie - shaking and pouting, his bug eyes wide and afraid, calling out for help.

Beneath the covers was the Great Emperor, Tiker-Lee, our vicious rat terrier, the most important animal to walk the grounds of the farm. Just ask him. He believes he is the most important thing known to human acknowledgment. Just as important as peanut butter is to jelly or eggs as essential to toast. And, when the sun drops down in the skies, Tiker-Lee is "placed" into the bed with me and my Wife.

That's right - PLACED!

Tiker-Lee does not walk into our bedroom and jumps onto his spot on the bed. Nope! Almost every night, Tara, with her ultimate love for this creature, will pick him up from the couch and CARRY him to the bed. If she doesn't, the little turd will lay on the couch - perched, more or less - head up, eyes wide open, an expression of WHAT THE HELL on his face, waiting to be picked up and lovingly placed on his "portion" of the mattress, which, basically, is anywhere he deems fit for his comfort.

This morning was a little different.

Almost all the dogs were in fear. A thunderstorm is brewing... and when it comes to thunderstorms, all courage seems to trickle out of the dogs like piss running down the side of your leg.

No, no. It is not funny, and I worry. The Fourth of July is nearing, and, good and traditional American that I am, I want to blow things up... shoot rockets off in the sky... celebrate in a loud &

patriotic way. Truth be told, I have a conscience. I worry about the dogs & what those loud explosions will do to the little - as well as big - four-legged pains in the asses.

For years, I would not fire a gun on my parent's property. They have a little dog, Buddy is his name, a little Maltese-bum-spoiled shit. The loud noises scare him, too. Even if I went to the gun cabinet and slid the door open to take a firearm out, Buddy would shake and hide.

(I have no memory of ever having him see me hold a rifle or shotgun, let alone firing a shot near this dog.)

There is no need to scare anything... if you can help it.

I just wish I could control the weather. My beagle dog is close to having a panic attack.

Bucky, the Beagle, is the worst when it comes to storms. He is also the *loudest*.

A Real Gem… & It's Still Going

For those of you that do not know, I'm a book dork. Collecting books and writing has been my passion for a very long time. Right across the room from where I am typing this, a bookshelf stands, loaded and bloated to capacity with volumes of varying styles. A complete set of the Complete Works of Mark Twain from the 1930s; an impressive (in my opinion) first edition collection of Hunter S. Thompson and Norman Mailer; of course, my first-ever love, Calvin and Hobbes books, published by Andrews and McMeel, written by Bill Waterson, and many other glorious books.

(Funny little piece of info about Mr. Waterson: he lives in Ohio. Born in Washington D. C., he moved to Chagrin Falls, OH, where he spent his time working on the much-loved comic strip. Just another Ohio boy turned writer. He's a guy I really look up to.)

I'm not sure when this wonderful, terrible addiction to words started… when the addiction "took over" to be classified as "active" addiction.

What I do know is that I was encouraged. For as long back as I can remember, reading was encouraged. Everyone from my mother, her father, Mern - even my best friend in 4th grade, his grandmother - pushed and bribed me to /read.

That's right - BRIBED!

Ms. S__, my best friend's grandmother, would pay me $2 for every book I would read. And here is the kicker: I would save that money up to buy *Calvin and Hobbes* books from the Scholastic Book Club catalog.

(I know, I know - I was hopeless! A junky to the printed word.)

There was also another program fueling my young and silly mind to read... and read... and read. And that program was called BOOK IT!

Do any of you guys remember to book it?

Let me refresh your memories:

"BOOK IT! is **a reading incentive program created by Pizza Hut in 1984**. We empower teachers and parents to motivate their students to read by providing Reading Award Certificates®, good for a free, one-topping Personal Pan Pizza® from Pizza Hut."

That is what I found online as a broad-stroke overview of the program. Now, from what I remember, I'm not sure on what day or why I was given a button/badge. It was explained to me that for each book I read, I'd be given a star sticker. These stickers would go on the badge - mine surrounding, almost blotting-out the whole damn image on the thing. After consuming so many stickers and so many titles, I got my own Personal Pan Pizza.

(For me, I always ordered cheese only.)

Now, there is more to it than that.

My mother would take us to Pizza Hut in Ravenna, on ST. RT. 59, during the summers. That was library day. And one of the great things about my mother - SHE ACTUALLY TALKED TO ME ABOUT WHAT I WAS READING! Later-on in life, she would talk to me about what kind of music I listened to. I don't consider those interactions as challenges. Not anymore. What it did was sharpen a weapon of debate and intellectually take into consideration things I was mentally ingesting.

Another great thing about those visits to Pizza Hut was the jukebox. My mother would give us some quarters from the bottom of her purse. We three kids would pick out some tunes and have a blast. (I would always choose a Guns N' Roses tune, usually "You Could Be Mine" because it was from the *Terminator 2* soundtrack.)

BOOK IT! is still around. That's right! Visit: https://www.bookitprogram.com/

"**Our traditional BOOK IT! The program is still going strong after 30+ years**. We welcome teachers to enroll their PreK through sixth-grade classrooms for our FREE reading incentive program."

Once again, that is something I found online. Hell, I would even go to schools to encourage the kids not only to read but also to get free pizza!

When I was growing up, ironically, one of my favorite cartoons, the main characters, ate pizza! The Teenage Mutant Ninja Turtles!

And, WOW! I have really embarrassed myself in this column at how big of a nerd I was when I was young. But, what the hell, this was a fun piece to write. It was even more fun to recognize that not all memories I have prior to November 23, 2021, were Horror Stories.

Childhood and life, as all of you know, have the perfect ability & capability to scar. No one can argue that. Scars can cover you all over your body; however, it is up to you how often you look at them. And it's up to you to look past them.

In the Summertime

Not that this should come as a shocker - the South is HOT! North Carolina is roasting today at 92 degrees, but it feels like 102. It makes me want to hug Willis Carrier, inventor of the modern-day air conditioner. Unfortunately, Carrier died in 1950.

Every state I have ever lived in seems to have extreme-heat summers. In Montana, for example, the heat was different. It was scorched – that is the best way I can describe it - and the sun seemed closer. Clear skies made it dangerous. No cloud cover, making your skin very, very vulnerable to burning. Very little humidity. Fires in the mountains were a problem. There was an actual "controlled" fire season to eliminate the danger of an actual wildfire. The sky would be filled with smoke... a constant smell of burning trees and grass suffused the air; you could taste it.

Houston, TX, was simply a different world. I mean that in every way: the weather, the people - everything. I dubbed it the Concrete Jungle. That's what it was to me. Temperatures would rise and rise, the atmosphere full of moisture - a concrete jungle/swamp, boiling the buildings and melting the mud. The heat of a big city with asphalt and concrete baking exacerbates the discomfort.

(I was happy to leave Texas weather.)

Ohio is just Ohio. It is hard to explain that state and its schizophrenic summers. Growing up there, there have been many changes over the years. So much so that I can't pinpoint a "typical" Ohio summer. I remember droughts. I remember summers when it seemed the rains would not stop, flooding the fields... stuck farm equipment... hairy-ass pin oak trees... deerflies, mosquitoes, and ticks galore.

Each state will have its summer struggles, discomforts, as well as joyous highlights. I can't wait for the highlights of my first North Carolina summer. 4th of July, fishing, and, hopefully, moving to a house where there are trees shading the area.

Just today, Tara suggested we go on a date to - hopefully - our new house. Technically, it's trespassing. (An *illegal* date. What fun! Does not take much twisting of my arm to do something a little questionable.) But we have money down on the place, so I don't see the harm in us bringing a bag of food and drinks to consume on the porch. We have done it before, about two weeks ago, and with no incident.

I just hope our luck continues. Then again, that is The Prayer of the Outlaw... we always are praying for luck to be on our side.

A Tall Man

My older brother is a large man. Much larger than me. Growing up, he was a big boy. Much bigger than me.

However, one out-of-the-ordinary day, I saw how slight he could be. There was a Reverend there – a man of the cloth, a man close to our family, who was greeting us, Mourners, on this special day.

It was calling hours for my aunt, who was not really my aunt. In fact, all I know is that she had a love interest in my great-uncle… it did not work out, but my grandmother kept her. When my grandmother decided you were family, by God, you were family.

Anyway, she was titled "Aunt," and that is the only important thing about it. – She was family. And she passed away.

It was her funeral.

My brother and I rode together to the ceremony. There he was, standing just inside the entrance, big and dry hand extended: the Holy Man. A man I respected, and, to this day, still respect, greeting all of us walking into the building.

Funny thing about that man: when he spoke to you, he spoke to you eye-to-eye.

When he spoke to me, we were eye-to-eye. I'm no runt! I stand a proud 6'1. My brother, on the other hand, has me licked by more than a few inches… and about fifty pounds.

Anyway, Reverend F__ shook my brother's hand. Faced him eye-to-eye. After the formal, masculine greeting, the Good Reverend stood up in all dignity.

In my eyes, he looked about SEVEN FEET TALL!

He had everyone in the room licked.

And that didn't mean a thing to him.

He was eye-to-eye with anyone.

Just Get in the Truck

For those of you not familiar with Ohio weather, specifically Cleveland, Ohio weather - it's damned dismal and grey and depressing.

That's how Ohio was when I left it. Most of my memories from that time have that feeling of foreboding desperation. GREY is the best way I can describe it. No. November of 2021 was not the worst part of my life. - Far from it. However, there was a downing, drowning suffocation in my life.

I was living in a semi-halfway house/rehab apartment complex... one that was not very "complex" but VERY restricting to the simple freedoms, such as going outside to smoke by yourself. There were random urine tests for drugs (yet drugs were very prevalent and easily available to any so seeking such indulgence) and alcohol.

This was not my first stay in such a place, but it was the most "normal" institution I'd lived in. And by “normal,” I mean I had my own apartment (complete with a roommate), was responsible for cooking my own meals and attending to my laundry; I had to pay rent and go to meetings - a structured environment that served me well... for the short amount of time I was there.

What was important - the MOST important thing for me - in those weeks was getting into the routine of daily writing and posting. There was a computer lab at the complex, and I had become very unaccustomed to having access to the internet. Almost a year! - A year without a computer! Throughout 2021, all the computer and internet access I had was for a brief stint in the summer. When I was allowed to go to the library, that was when I got onto a computer. That is also when I picked up writing and posting things on WordPress.

It was scattered, at best. I'm not even sure how many columns I posted during that time...?

What I do know is that was the beginning of what has become a glowing positive and educational part of my life. It's what kept pushing me to what became my first book, *Notes from the Night Desk: Volume 1*. And my second book is soon to follow as long as I keep up the daily work going. This has been an inspiration to keep on going... not that I would STOP writing. It's an inspiration to keep going, believing my dream is possible: to be a published author. To have the books in my hands, I have written... and that improvement in a life of major screwups is possible.

Before I moved into the COMPLEX, writing was a daily thing. I wrote on an old typewriter, that is. And in notebooks. I still have many pages of daily life in a sober house in West Cleveland stacked up in a file tray, unpublished and raw. I don't know if I'm ever going to do anything with them. File them away, of course, but they seem irrelevant to my life now in North Carolina.

And that is what this piece is about North Carolina and how I got here.

After two weeks of living in the COMPLEX - two weeks of writing and posting daily - I got a message from a woman.

"You intrigue me," said the message. And this intrigued me. For I knew the sender of the message... yet I had not been in contact with her in 20 YEARS!

She was a girl that lived up the road from me. We rode the same bus to school.

What the hell was she doing getting in touch with me?

I asked if she had been reading my column. When she replied YES, I was very happy. I was also very cautious.

Well, what more can I say about this? We started messaging and calling one another daily. She became just as important in my daily life as writing. We actually SPOKE to one another. There was no pretense. Nothing that made my mind uneasy.

She mentioned hiring some help for her farm. And, boy, did I jump on that.

There are a great number of reasons I had for wanting to leave Ohio - leave Cleveland and all the bad that is found in any city. It was the closeness that really sealed the deal for me, though. Tara and I became close. We talked in a way and about things I found comforting. We wanted to be around one another. – All the time.

So, on November 23, 2021, a 2021 Chevy Silverado pulled into the parking lot of the drug and booze COMPLEX. I had two bags, a box, and a lamp. It was a cold and miserable morning. My nerves were shot, and my anxiety was through the roof - IT WAS HAPPENING! I was going to start my New Life.

I was moving South. Moved to the South with a woman that seemed to like me very much.

JUST GET IN THE TRUCK, my mind shouted.

And I did.

No regrets.

This past year has had some of the best days I've had in... well... ever.

Tara and I are never far from each other. We have not spent a single night away from each other. People can call it codependency. Some people consider that a negative thing. If what I have now is considered negative, I wish I could have put those people in my boots two years ago, and then their definition of negative would have been drastically altered.

From Ohio all the way to North Carolina, we talked and talked about whatever came to mind. There was no way I could have known how much we would come to care for one another. I think she kind of found my meager belongings offensive, as in, at 37, I barely had any.

She took me shopping. She showed things to do on the farm. She encouraged me to write! We have moved to three different spots on the farm, and as I am typing this, we bought our Dream Home. A wonderful log cabin – Our Home!

The night before she picked me up from Cleveland, she asked what it was I wanted in life. I said I wanted a home. Not only do I now have a home, I have a life.

I married that woman on April 01, 2022, in Ashville, NC.

You'll read more about that as you go along in this book.

Being Put in Place

Tonight, it is an odd night. I am reminded of how wonderful of a woman I have. Now, that is not odd, but the fact she placed this laptop on my lap, telling me to write is odd.

In the past, I would write. Just to write was like a gift from God. I would just do it.

In the past month, a month where time seems to be going as fast as a second, I have not been as creatively productive as I usually am.

However, the wonderful woman who seems to feel strongly for me puts the writing machine on my lap to communicate with you wonderful Readers. Reminding me I have a job to do.

There are two dogs to my right, sleeping as lazy dogs will do. It has been an amazing month of moving and dogs and fights and wonders. An amazing life!

If any of you have ever felt the hell-bent fury of concrete loneliness to find yourself in a home with a family, you might know where I am coming from. A family of dogs... a family being brought back together, of making new members of that family.

I want this to be a short and sweet piece.

What makes this so wonderful is the assistance/insistence my girl put on me doing it.

She knows this is what I love doing - she knows it is who I am.

True, true: I have been lax in my responsibility as a writer. Now, let us all take into consideration EVERYTHING that has been going on. Everything from moving, selling, and wrangling up pigs to trying to fulfill legal, parental obligations... my goodness, my life has never been so busy, wonderful, and complete.

And there are all of you out there.

All of you, reading this.

And, oh, how I appreciate and love you.

Next month is going to be a killer-hell-woven-freakish time for my writing.

My girl will be starting a new job. She will be working at home with me. Making sure I did what I was supposed to be doing. And, yes, I need that kind of support and encouragement.

Why not admit it?

For all of you out there, everyone should have such care and encouragement.

Volume 1

This morning at roughly 8:54 am, I gave final approval for the book to be uploaded and available for purchase on Amazon.com.

My Project Manager informed me it may take 72 hours before it can be purchased. But the important thing is the bulk of the work is done. Of course, there is little concept of promoting it. Having people buy it. However, the important thing for me was to finish it. This was a personal, hellish & glorious venture into trying to obtain some sense of confirmation.

I wanted - almost NEEDED - to feel a certain sense of accomplishment at what I have been working on for years.

Not the book.

But writing.

Yes, yes, yes: the first piece in *Notes from the Night Desk* was inspired by a conversation I had with my sister back in 2013, but the bulk of this bastard-of-a-book was written in four or five months. Every day, I was on this keyboard, thinking of what could possibly be interesting enough to share with others...and who would care?

Well, on the phone with my mother yesterday, I got a little bit of confirmation. My mother will go out for tea with some of her oldest & dearest friends. It was mentioned by one of these ladies that, after reading one of my columns, it made her laugh out loud.

That made my heart happy.

I made someone laugh.

Sometimes, that is what a person should have before having to mop the floor, pay the bills, and clock into a job they may not like... hell, sometimes a laugh can stop a person from doing something really stupid and permanent.

That is a great feeling. It is a great form of confirmation that I am writing in a way that can reach a person.

No! I'm no Hemingway or King or Faulkner or Twain. - What I can be - and what I can write about - is myself, in a style that I find fun to work with.

Notes are not the beginning. It's the evolution of something I started back when I was a kid.

I remember the first story I wrote. I don't remember how old I was. I know I did it because my uncle had given my family a word processor. The story was about a mountain man. He hunted & wrestled grizzly bears. And that was what I wanted to be and do.

I have wrestled some grizzlies in my day. Not the type you see working their way up the side of an Alaskan mountain, or the type that has been skinned and made into a rug... but they were grizzlies, nonetheless. And just as vicious.

My proof copy of *Notes from the Night Desk* will be in my hands by Tuesday, April 05, 2022.

My first book!

And I can honestly say I have been putting off a good deal of things, waiting for that damn thing to be in my hands.

I thank all of you for continuing to read and support me these months. All the kind messages of encouragement have been a blessing... and an eye-opening lesson about how good people can be.

Breakfast With Hemingway

On the fourth floor of the Cambria Hotel in downtown Ashville, North Carolina, there is a wonderful, exotic, and important place to visit: Hemingway's Cuba Restaurant & Bar.

That is where my wife and I had our wedding dinner. It's also a place I visited four times during our three-day stay at the hotel. Not just because of great food and great service and the like... it had to do with the feeling Hemingway was, in some way, with me.

It was Ernest Hemingway that convinced me I was a writer.

Back in high school, way before I ever thought I would make anything out of all the time I spent writing in notebooks, I read Hemingway's *A Moveable Feast*. In those pages, Good Ol' Papa described going to cafes in Paris with a notebook and pencil and writing stories.

Hell, I thought, *that is what I do. I must be a writer.*

It only seems too appropriate that I would have my wedding night there.

I return the following morning to get my glorious bride some breakfast and return the same night to have dinner. At that point, I was so addicted to the place I was making the newly-blushing-bride blush, not because of the "you-know-whats" but because I was hell-bent on stealing the bust of Hemingway on display at the restaurant.

(What can I say? They refused to sell it to me.)

I also had my GONZO pendant dangling from my neck. Hemingway and Dr. Thompson with me...

I was in a major energetic-inspired mood this past weekend. Not only did I get married, but my first book was also made available on Amazon.com. *Notes from the Night Desk: Volume 1* is available both on Kindle and in paperback.

By the time my bride and I were back from Ashville, my Author's Copy was waiting for me in the mail.

Now, back at home, in Taylorsville, I'm trying to come to grips with the magnitude of what has happened in the past month... hell, just this year alone, so much has happened.

Tara and I are just getting situated and comfortable in the new home one week. The next we are getting married. My aunts were there. Two beautiful pillars that reminded me of family will always be there. And that Tara is now part of that family means a great deal to me. I know my aunt Brenda & my aunt Cheryl will be there for her, no matter the condition. I know they would be there for me. - They were there for me on those special days. My invitation to them was very purposeful, meaning that they were specially selected for reasons I can't put my finger on. Just knowing they have cared so much through the years means a great deal.

All of the congratulations from so many people on social media - on the wedding and the book - have made my heart very happy. I thank all of you. Thank you to my parents, sister, Aunt Brenda and Aunt Cheryl, Prof. William McKeen, and Ron Whitehead, to the wonderful friends that were there at the ceremony and for their support & laughter.

And most of all, To Tara Locke!

It has been one hard-hitting year thus far. I imagine it won't slow down.

This coming weekend, we are going to be back on the road, heading up to Ohio and visiting family. Oddly, I posted on Facebook that I will be eating dinner at the Alliance, OH, Applebee's if anyone wants a signed copy of *Notes*.

More people responded than I had anticipated.

Facing people in such a situation is daunting to me. I never fully trust myself when in public, the center of attention, and having to feel the need to "perform." I know it's not a big deal, but, in a way, it is.

An Idea

It's funny what your mind can come up with when you have a few moments to yourself. The kind of responses you get from a brilliant idea is even funnier.

I'm not sure why the idea came to me while I was in the bathroom. But that was my brilliant idea: *we need oranges!*

"We can't have Free Radicals loose, damaging DNA," my wife said in response.

"Good point," I said. Not knowing what in the hell she was talking about.

She continued, "Defense mechanisms against Free Radicals... Antioxidants like Vitamin C..."

"From the oranges," I shouted, trying to understand where the hell this conversation was going.

"... cells to mutate cancer," my wife went on.

I was in over my head.

"Metabolism of acetaminophen produces Free Radicals…"

God, she is sexy, I thought, as she carried on and on about medical facts and jargon far beyond my comprehension. It was late Friday night. Tara was in one of my t-shirts, snuggled up in a blanket. Dogs and a puppy surround her.

I'm not sure why I suddenly thought we needed oranges. I think I just had a craving that, hence, turned into a medical lecture.

And I have no room to complain. I'm sure she has heard more about American Literature and writing being with me than she ever wanted to know.

Everyone has their own thing.

My wife... her brilliance is in so many categories I can't keep up. Medicine, however, is her true forte. Love and care of animals come in a strong second.

(No wonder she married me: I'm a challenge when it comes to health.)

Last night, I was reminded of how wonderful she is. And how much she helps people - including ME!

This morning, while at the gas station, I bought an orange.

Ghosts In Boxes

Tara and I returned from Ohio this past week. We were visiting family, celebrating the fact we got married. We did not return empty-handed. There were boxes and boxes and boxes of things I thought I had lost, given back to me by my parents.

Boxes of books, movie memorabilia, photographs. The most important thing is a very large black, heavy-plastic toolbox (the type you affix to the bed of a truck) full of my early writing. Years and years' worth of writing. There is even a thin brown leather journal filled with terrible poems I wrote when I was 16.

Rare magazines with articles and dedications to Hunter S. Thompson were found. An old portable typewriter from World War 2, a stuffed Friedrich Nietzsche.

Many things from former lives...

I did not expect to be impacted the way that I was. Specifically when I opened one of the plastic boxes. Beneath the books and horror movie memorabilia, there was an old Bible - My grandfather's Bible. I was in a hotel room with Tara, about to go to a restaurant to sign some books, close to a panic attack - I don't do well in crowds - and there was this old Bible.

Now, I am not a religious son-of-a-bitch, but I know that book meant something to my grandfather. And it was something he left to me when he died.

Seeing it calmed me down.

It also made me very emotional. It was a treasure I had thought was lost... burned up with so many of my other possessions from the past.

The past will always catch up to you whether you want it to or not. But sometimes, sometimes, it can be a comfort.

The point is, never assume how you will feel when that finger taps you on the shoulder to remind you of where you have been... and the people that were included.

April Afternoon

To settle down after a hard, hellish, and stressful stretch is not an easy thing to do. There needs to be a lot of rest. A lot of support. A lot of things that don't need to be mentioned. That's how it is for me, anyway. Rejuvenation is a *must*.

Routine is such an important thing. Just ask my son. And through all the chaos, all the mess, something resembling a sane & orderly life should emerge.

Right now, I am looking for that - that balance. As good as things are, there are still those moments throughout the day where things are... well... difficult. Yes, I know, EVERYONE has those moments. They have those moments every day, probably. Me and my wife are no exception.

Things need to be done. On the farm. In our home. In our professions. Now, I am a man of leisure. That is not to say I am a lazy man. In the past month, more external stimuli have come my way than what I am used to. So much family has been seen. So many changes and evolutions are happening... it is so hard to believe from where I was a year ago to where I am out now.

There are a few things I could do to change and dismiss and better the daily situations.

Let me tell you, dear Reader, a story: My eyes opened in the middle of the night. I heard something moving around in the closet. Tara was warm and asleep next to me. She had not woken. I got up, reaching for the .357 mag next to me. There was a struggle. I could tell something was happening in the closet. Opening the door, I saw two people.

How the hell did they get in?

Past the dogs!

Past me!

I opened the door, and there they were: Jim Morrison between Amy Winehouse's legs, kissing her hard on the mouth. He turned, looked at me, and smiled that famous smile that seemed to impart a faux sense of surprise and entertainment.

Jim's wise words to me: "You need to wake up, man."

I looked down at the floor. There were hundreds of pages - MY hand-written and type-written pages - on the floor. Jim's boots stepping and ripping them up from the excited tussle.

I needed to get back to bed.

And I'm sure Jim and Amy wanted to get back to doing what they were doing.

When I woke up, I thought about all those pages... pages I had been reading the day before.

They are all here with me now, waiting to be polished and properly placed.

I'm not sure if dreams mean anything - anything as in that cosmic sense of truth and reality. If they do, I'm sure Jim is having a good time.

Hopefully, Ms. Winehouse is, too.

Stronger & Stronger

EASTER MORNING

As I wrote in another column, I'm not a religious s.o.b. However, because of my wife and her affiliations, we will watch a live-feed sermon from time to time.

Preacher Jamie Steele, a man I had the privilege to meet at the local Wal-Mart, spoke of the Resurrection this morning.

Of course, I thought. It's Easter. A Christian sermon on Easter... what else is there to speak of? The Resurrection - yes, yes. The topic of the day, as it should be.

Near the end of the sermon, Preacher Jamie switched gears. He started speaking about hope. He told a story about visiting a prison. He asked the Warden who the most dangerous man in the prison was.

"That's simple, Preacher," the Warder said. "The man that has no hope."

Talk about a punch-to-my-face recall of days gone by. Bad days. Dozens of memories, hundreds, and hundreds of days came back to me. Days when I had no hope - NONE whatsoever. Not only would I do or say anything, but there was no conscience in my actions because there was NOTHING, nor nobody, I was wanting. No hope for ANY future.

I drifted and floundered that way for years.

When did I get Hope back? How did I get it back?

I'm not sure. And it's not important. What is important is I grabbed it, and I will not let it go. Sure, there are struggles every day. It would be false and wishful thinking not to acknowledge that. But to hope and to dream is SO important. Trust me, dear Reader - it is very important! Something needs to keep the heart pumping and the brain working.

Every night, I go to bed with the hope and belief things will grow stronger and stronger in what is most precious in my heart.

Hope. What a silly and essential thing. At least for me. Yes, yes - it can be dangerous. The letdown, the down stroke from disappointment is ever a threat... but take it from me - it's worth it!

Yesterday, Morning

Not all things can be good ALL the time. Even at this secluded location, I call home, family strife still arises, and political unrest still takes in and exhales deep, poisonous breaths over the television, computer, and cellphone screens. Couples still squabble over the mundane annoyances of life with one another... hardship has a never-ending, iron-clad grip on everyone, everywhere, in some form or another.

It could be the third grader that forgot his or her lunch on the kitchen counter, with no time to remember to grab it; a smalltown businessman having to close-up shop because there were not enough profits to keep operating; a dog with cancer uncontrollably pissing itself beneath a master bed... people lacking emotional intellect to help them understand that problems are not just designed for them, and let's not forget the world's dwindling grasp on being patient.

And, some days, I feel like all those things are me. - That they affect me, frustrate and limit my serenity, and I'm just another intolerant asshole not mature enough to handle the frustration - the drama!

Then, things like this morning's farm chores changed all that.

I woke up in a good mood. I fried an egg, shredded some gouda cheese, and then sprinkled the cheese over the steaming egg, placing it in a dinner roll for a nice breakfast sandwich. To wash it down, a cold 24 oz Starbucks coffee/energy drink.

It was a fabulous breakfast. And the sun was shining. A little chill in the air, but it brought about a bright and good feeling.

I wanted to get the pigs, chickens, and goats fed early. Before Tara was home from work, she had been getting up at three in the morning to work her second job at a clinic specializing in addiction care. She usually comes back early in the afternoon.

If anything, a partner should think of ways to make their loved one's life easier.

That is what I thought about while driving the side-by-side ATV to the goat and pig pens. We can always find a way to make someone's day a little better. *Action* is the keyword in this Rant of mine. To make something better - to make something worse - SOMETHING must be done. Sometimes, doing nothing is something that will ruin your day and your family's day...

Start off with a good breakfast. It helped me this morning. Then, graduate to doing something for someone other than yourself.

They Were All Wonderful

All of them. All the typewriters I have owned, borrowed, and used throughout the years have been special. Each one participated in at least ONE GOOD sentence typed up by me. There is something about the paper rolling on and on and on, filling up with words. Didn't matter if it was a poem, or a letter, or a stack of pages of a short story, or a much longer manuscript that could be the treasure starting a Novel.

It was all about the noise. It was all about the page riding and inching along and the "CLING" telling you it's time to whip the bastard back into fresh attention. Time to start another line.

I write about this - writing on typewriters - at least twice every six months. Right now, on a shelf above where I type my nightly work on a laptop, sits an Underwood No. 5. And it is a machine that has been through many battles!

(I think the model is from 1919, but I have seen similar Underwoods from the 1930s. They look the same.)

This typewriter is rusted. Every piece of it needs to be refurbished and worked upon. All the keys are still intact, though. It is a great "art piece" or conversation starter to have. But it is a non-functioning item.

A few days after Christmas, Tara and I traveled from North Carolina to Ohio to visit my family. Gifts were exchanged, of course, but it was my father who took me by the shoulder and pointed to a large object wrapped in a garbage bag.

There it was: an old typewriter - that very same Underwood I had in a house years ago - sitting on a table. We were at my aunt's house. I knew the damn machine looked familiar, but for the life of me, I do not remember when or where I got it.

What I DO know is it was a Treasure from my life before. My life before all the BAD things started to get out of control.

Whenever I write with a typewriter - depending on the typewriter - I think of Hemingway, e. e., cummings, Bukowski, Jack Kerouac, and all the beautiful Beats... if it's an I. B. M. Selectric,

I think of Dr. Hunter S. Thompson. I think of Jack Nicholson's portrayal of Jack Torrance going insane, working on an Adler Tippa S typewriter in Stanley Kubrick's *The Shining*; James Caan as Paul Sheldon, forced to write on a "fifty-pound clunker" Royal 10 typewriter in *Misery*.

A *flâneur*, that was me, returning to my room or office after a day or night out, watching the world and the people of the world around me, and then sitting in front of one of those big, ugly, metal and noisy machines - it was a GREAT time, my early-to-mid-twenties. The time I spent with those magical, archaic keys... filling up the pages... the pages piling up, face down to the right of me on the desk or table.

Most of the writing I did a year ago was on a Smith-Corona from the 1960s. Later in the year, an electric AT&T 6100 replaced the old manual girl. I would switch from one machine to the other, depending on my mood. Some of the time, I would be working on two things at the same time, darting from one end of the small room to the other, making much noise, and freaking out some of my housemates.

Most of the time, though, I stuck with the old manual.

Most of those pages are with me, untouched and almost ignored.

Why am I waiting - or, more accurately, reluctant - to type them up on the computer to post?

I guess a lot of it has to do with the fact they were written somewhere else by a person I outgrew. Growing (and growing FAST) has become a big part of my move to North Carolina.

My time spent in Cleveland, for the most part, served me well. But like so many of those places, a person can outgrow them... they can *evolve*. - Then it's time to move on.

I made the choice - the *gamble* - of leaving Cleveland for the golden-warm promises of a better life in the South with a person I had faith in. I had to believe what was said in our phone conversations was sincere. That there was a home waiting for me.

So far, North Carolina is greater than I imagined. Greater than any fantasy I conjured up while curled up on some bed that was not mine or in some abandoned building in Ohio. And I'm close to someone. Someone who trusts and believes in me.

Still, I have these typewriter pages... and something SHOULD be done with them. I have the Underwood No.5 above me, reminding me of a piece of my past that was not ALL gloom and doomed days.

Maybe I will go through those pages and find something nice. When I do, I'll share it.

A Few Days Away

(Dedicated to Dan & Kim Locke.)

Tara's attention is divided between her work, looking at her laptop one moment and the next second, she'll be on the phone with a client/patient. She is a general nurse practitioner, performing her magic in telehealth appointments for a virtual clinic based in New York.

I'm not even going to pretend to know how to write about what she does professionally. I respect her and her field of work too much to butcher the goodness with bad writing.

What I DO know, and what I am qualified to write about, is what she does when she is off the computer and when the phone is not stitched to her hand. Today, it would be preparing us for a trip to Asheville, NC. That is my birthday present: two nights in a nice hotel in what she says is the "freakiest" city in the state.

"But, compared to you, it's tame," she said with a smile and a laugh.

I like the idea.

As a Professional Freak, I am curious to see what Asheville, North Carolina has in its buffet. Not that I'll be seeking out any kind of wildness. Don't really need to. My average day, temperament, and actions satisfy me just fine.

About the wildest thing I'm going to do is find a used bookstore. Well, maybe there will be some "other" wild things I'm going to do. But I want to keep this column as PG... PG-13, max, as possible.

(My little nieces might read these pieces of mine one day. The last thing they need to read about is their Uncle Andy-Boo's adult exploits.)

Laundry, of course, needs to be done.

Reservations need to be made, I think. When I was talking to the helpful Hotel front desk lady on the phone a few days ago, she informed me it would be wise to make dinner reservations. Sadly, it has nothing to do with the restaurants being fancy or busy. It's the off-season in Ashville. Nothing much is going on. It's not the time for tourists. Reservations should be made because of low staffing.

This is going to be the first birthday I have *truly* celebrated, in a traditional sense, in years.

Two years in a row, "celebrating" my birthday was going to a restaurant with my parents... after they picked me up from two different hospitals, in two different cities, over the same thing. On some of the other birthdays, I was just lost. Or in institutions or rehab. Birthdays are wasted without phone calls or cards.

Anger.

Frustration.

Disappointing those that I love most.

That is what I turned My Special Day into.

Cynically, it's just another day. That is how I justified it. And why and who in the world wants to celebrate the day I came into the world? The last thing I needed, at the time, was a reminder I was somewhat attached to a world I felt discarded from.

It's supposed to be different this year. And, oh, things have been different and wonderful and new and surprising.

Hopefully, all those things will make it to 12:00 am January 23, 2022.

Hopefully, it will go beyond that. I hope everyone can feel and witness some of the joys and miracles I have been gifted with over the past few months.

A Question Answered

I asked my wife, Tara, when I should get started on *Volume 2*, my second book. She said YESTERDAY! That was a pretty good damn answer.

What are my goals for the second volume of autobiographical columns? Well, hell, I guess just sharing with you, good Readers, about the daily life of a piss-poor, pain-in-the-ass person hell-bent on being a writer. And some of the struggles a good-loving woman has to hold the hand of such a person.

See: I would be nowhere & nothing without my wife. If anything, she kicks my ass when it needs to be kicked, putting me in my rightful potted place. I can be a stubborn man. A lecherous & idle man, always with ideas brewing in mind - but needing a good kick-in-the-ass to sit at the fucking laptop or typer to get it down.

That is what she did today. And it is something she does for me every day.

What a woman!

The dogs get to have their wonderful life being dogs, sleeping & eating at their own pace... Oh, what it would be like to be a dog!

I think that is what I would like to be reincarnated as if such a thing exists...

Back to the next book: There are going to be some pieces of writing I will not publish on this blog. For a couple of reasons, I guess.

1. They are going to be pieces I wrote from my past. I was an altogether DIFFERENT writer back in my 20s. (Hopefully, I won't embarrass myself too much.)
2. Some of the content in the next book will be for ADULTS ONLY!
3. I would like Vol. 2 to be something different. Not just recycled columns from this website. Either way, I'm sure all of us are going to be entertained.

A Torch Burning at Both Ends

The files are piling and piling up... with so many files still needing editing and organizing. I look at the stacks of pages and the files on the computer, and I can't help but think of my mother and what she must be going through... what she must be thinking.

My mother is on the verge of being a great writer. The hell with that - she is A GOOD WRITER. But she is getting to that period when a writer goes through a tough, essential spot. It is when the ideas and the pages and the files start filling up the empty spaces of your home, not to mention your mind. And, of course, the question: WHAT THE HELL DO I DO WITH IT ALL?

What a fabulous, frustrating time, dealing with time, creativity, and what to do next. How often should you work on one project, finish the first one, pick out the Author's Photo, write the Author's Description, and all that fun stuff, which is (for me) not that much fun, but it is important? When you have hit the point of your journey when the "cosmetic" approach is the next most comprehensible venture to tackle, you should feel a *little* apprehension. Don't ask why I feel and think that way; I just do.

I talked with my mother today about several of her writing projects. She seemed nervous. And it is understandable. There is a great deal of self-pressure involved in creative endeavors.

Am I doing this the right way?

What am I doing wrong?

Will others like it?

So many, so many annoying questions. Hard questions one can't help asking themselves.

Some things are in your blood. With some of the people in my family, telling stories - sharing stories - is something we do. Just like any other form of knowledge, the desire to share it comes naturally to some… when the knowledge happens to be delivered in an entertaining and comedic way, well – all the reason more to open your trap and gab about it.

Listening to my grandfather, my uncles, my mother, and my grandmother - my great-grandmother, even - I received an education in the importance of words and stories and the passing down of them to a generation of the next pair of ears.

So, listen to your elders. Even if you don't agree with them, it will/may define who you are. I know it did for me.

My mother is a writer. Same as me. We tell our stories, our beliefs, our past, our doubts, and our passions. Our accomplishments and our failures are verbalized or written down as common Moral Tales of living life during the end of the twentieth century and the beginning of the twenty-first.

And to be a good son and (laugh at myself) *teacher*, I am happy to contribute any help which will further the trip along the destination my mother wishes to travel.

If ever you have a family member needing an open ear to fill, let them fill it. It feels better and sounds better than a hole's echo from the minds of hell.

A Good Job

Sitting and glaring at the laptop screen, a typewriter's keys, or tapping a pen on the blank page of a notebook or legal pad can cause a real conundrum – a real pain-in-the-ass process of knowing I can do it… but not right at that moment. The WANT is there, and the need is there; however, at times, there is a reluctance... a lazy, reluctant bone not working the weight properly to work or pump the blood to its appropriate destination for motion. A person can come up with dozens of reasons *not* to write.

The dishes need to be washed, the laundry folded, or the bed needs to be made, phone calls need to be dialed and answered, the dogs need to be put out - ANYTHING to not start the WORK of the WORD.

And that is the hard part: starting the process.

Once the fingers start to work their magic, you are golden. Words start to form and flow in river-like speeds of soft, tender motion, to a violent force over rocks and hurdles of the mind and anything and everything you define as SELF.

You - I - just have to begin.

There is nothing like it. Once the magic begins, the high starts. Even better, when the stack of pages starts to get thick, when you can physically touch the work, flip through amazing sheets filled with words... it's a wonderful thing.

It is a wonder I waited so long to begin such an amazing event.

(What the heck was I waiting for?}

There is no proper or correct answer to that question.

Waiting for the moment... now that, that is *really hard.*

I always imagined being voracious, especially as a young man. Pages and pages and pages typed up every day, every night. Entire novels are written in ten or twelve days. However, that is not the case. However, I have written three small books in a year.

That is something.

One of the best jobs - if not THE BEST job - I ever had had nothing to do with the paycheck. No. No. Writing is what I do now, and some money has been made. Receiving a regular paycheck for my efforts has not happened. Not yet, anyway.

The bookstores – Follett, Kent State University, and Trumbull Branch – had a lot of downtime, which made it perfect for a college job. I would juggle my class schedule, my work schedule, and my private life in perfect harmony. And with all that downtime, I could focus on coursework, leaving my midnight hours dedicated to writing.

Of course, there is the subject of my helpless addiction - to BOOKS!

I worked in a bookstore, for heaven's sake. That would be like giving a junky a job at a pharmacy, handing them the keys to the narcotics kingdom. The key to the goods, and expecting there not to be a cataclysmic outcome of a product being devoured, is absurd.

And I devoured. And devoured some more.

Everything from biology texts to 1960s radical political movements, classical English poetry, and German philosophy - I devoured it all as much as I could. No book was safe from my Hunter's Eye. In pursuit of becoming better, doing what I love: reading. Learning the Craft of writing is so much different than having the passion.

It's so much harder, as well.

Throughout the years, I have had many types of jobs. Everything from fry cook, dishwasher, landscaper, staff writer for a newspaper - and photographer at a daily newspaper - but no other job made me feel more at home than the bookstore.

The bookstore allowed me to be ME! In every way, you could imagine… for a nineteen-year-old kid, at least.

I worked there for three years. Full-time and part-time – seasonal, as well as a concrete position.

Family can be found in many places. At jobs, in a house, volunteering, and wandering the streets. – Family is everywhere. It's all about how important and memorable the people you surround yourself with define a brother, aunt, or sister.

Blood really is not the determining factor.

It's important.

Pyr Pressure

Something I must deal with every morning, every day, and every night is Pyr Pressure. Not the type you might be thinking of... not the type where I am offered something, and others are around me saying, "Do it. Do it. Everyone else is doing it."

No. No.

I'm talking about something more annoying. The Great Pyrenees - the great white dog - big and gentle and kind and protective, has a tick. A personality trait, if you will, of sitting right before you, their big brown/black eyes focused on you, their *MONTE DE PIEDAD* of whatever plight you might have frozen on their faces, and they want to PAW you.

That's right: PAW YOU!

They will sit their big, white, and fluffy rumps down and raise a leg, put all their weight and pressure into one foot, and then crush you with it.

It does not matter how many times you swat or pry the Pyr's paw away, and they continue to raise that leg and paw you up, put it back on you, and give you that pitiful look. A look that shouts: "I want to help."

They truly are wonderful creatures, full of majestic qualities. Their most endearing element would be their want for affection and attention.

In all of the five Pyres my wife and I share our life with, not one of them seems to have a mean bone in their bodies, and they want more than anything to cuddle, be petted... be acknowledged.

If it's not the paw they get you with, it's their noses. They use those noses as spades, digging into your side, your underarms, shifting through parts of our bodies, looking for attention. They will bury those big noses in whatever section of our body is the most vulnerable and then lift with all their might, causing us to shift and sift our weight around in affectionate play.

It can be cute.

It can also be a pain in the pancreas and an annoying distraction.

I can be writing – caught and lost in the moment, deep in thought, and in a hurry to get the words down - next thing you know, I have a nose under my arm, lifting my hand from the keyboard. And those faces are just looking and pleading and begging…

They are a breed demanding attention. (Not that I can blame them. I am demanding attention, as well.) But they are also the type of dog that knows your emotions, your physical well-being, and the difficult and, sometimes, futile relations you can have with another individual.

Hailing from France, the Great Pyrenees, also known to the French by the title PATOU, were mountain livestock guardian dogs from the French side of the Pyrenees mountains, not far from Spain.

And what a dog breed they are - a force in affection and protection and slobber. Big eyes look at you very sincerely, questioning what to do next, where to be, and/or when they will get their next treat or when their ears will be rubbed.

HOW CAN I PLEASE YOU? Those faces seem to ask.

As much of a pain in the ass they are, they are also the most loving of animals. Our home just would not seem like HOME without the great white beasts. Our hundred-pound lapdogs dig away at us, demanding our attention.

At first, when I moved down south to be with Tara, she warned me she had a "dog problem."

NO PROBLEM, I thought to myself. The dog issue would be her issue, not mine. The dogs would be HER thing. Not mine. And then, without warning, time went by, and they became MY dogs, as well.

Why not? I ask myself one summer afternoon. I have dogs now, too.

What is so wrong with that?

Well, I can name a few problems that have presented themselves to me; however, the perks very much outweigh the cons.

Nothing I can think of could make me want to abandon the fluffy Polar Wolves.

Not a freaking thing.

So, with ten dogs, a wife, two miniature goats, a dozen chickens, fish in a large fish habitat in the living room...

I'm creeping on my second year here, in the First in Flight state, my dream home acquired... and it is filled with animals.

I would not have it any other way.

My wife, my pack of dogs, wildlife, and farm life...

What does one do when all of their wishes have been granted?

When I discover the answer to that question, I'll be sure to share it with you good people.

Another Short Piece

It is that time again. The time when I should be finishing a manuscript to turn into my Project Manager/Publisher. And, once again, I am putting off working - being lazy, NOT WANTING to transfer files from one device to another. Not to mention NOT wanting to edit the new content I have been working on for the past few weeks.

Whoever thought the last lap was the easiest must have been insane, on drugs, or just totally fucked in the brain, like a marathon runner telling themselves: just one more mile, when, realistically, there is twenty miles to push through before the finish line.

Not like I don't want a new book to come out.

I really do want another title available for readers... for myself... to see what the bastard will look and feel like when it is published. There is nothing like it: seeing and holding your new book for the first time. It feels like a victory. As if the hardest battle - the muddiest and bloodiest battle you can imagine has been won, the flag has been speared into the dirt.

For what it is worth, it has been worth it. All the frustrations of getting started, getting through the first draft, thinking of how to end it - it has all been worth it. And the book WILL be finished soon – IT WILL BE FINISHED!

The End of Brotherhood

Not so long ago - maybe longer than I wish to remember - I had two brothers. An older brother. Two years older than me. And a younger brother... how much younger…? I am not sure.

For the life of me, I'm not sure what happened - what I did that was so horrendous - that the relationship between me and my brothers withered and died.

Not just died in a quick, gunshot flash. No. This thing is still going on. It's festering and rotting and turning uglier and uglier, distilling into an acidic poison.

It must have been something godawful. Whatever it was I did. My brothers are not the type to forgive, admit when they have done wrong, nor admit when someone else is right. They have to be right. They always have to feel they are in the right. And nothing – and I mean NOTHING – will deter them from their convictions.

(In a way, there is honor in that. However, it makes them impossible individuals to tolerate. I can admit when I am wrong. And I feel no threat in another person knowing something I have no knowledge of.)

Something that does not matter to me but means something to them: an uncontrollable urge to continue with an already dead reality of the "family" land staying in the family. Hell, they don't even have homes in their names, no land, unpaid for vehicles, and working slave hours… what could have been the best of times (at the age in question) my older brother was working sixteen-hour shifts, six days out of the week. Three or more so hours A DAY and NIGHT just driving to and from work!

Pathetic and narrow-minded and judgmental as they are, with their hatred and desperate need to hold onto a grudge, they seem incapable of focusing perception on themselves and their actions.

And, really, what is the point in doing so? To prove a point? To show how unforgiving certain people are? To show their moral "superiority" over an individual with a legitimate, medically acknowledged disease?

That does not seem very moral to me.

Hell, I learned their lesson – the lesson of unforgeability - a long time ago... back when I was ten, wearing short pants, and thinking Rambo was a real person Some people don't know whatever lesson they are trying to impart has been taught over and over and over again, throughout the years of one's life – a constant in being taught how deplorable people can treat one another.

They simply perpetuate the worst and close-minded essence of the human condition.

One of things that separate us from the beasts is the ability to grow, forgive, and learn.

My brothers are no better than cheap, over-played poorly written political oratory, desperately trying to convey a message devoid of meaning, lacking in conviction.

Flames in the Dark

Sitting outside. It's nighttime. I'm waiting for my wife to come home from work. Sitting on a lawn chair, smoking, looking at the fire burning in two lamps, the flames from the firepit, and some tiki torches burning their fuel - their scented smoke suffusing the North Carolina air...

It is all so good, I think as I flick ash into black, plastic tray on the porch table.

How did I - me of all people - get to be where I am at. Not only do I have the woman of my dreams, but we get to share this... this Southern paradise, together, as a unit. A unit dedicated to itself, and what we have wished so much for, for so long.

The weather is getting warmer. Tree frogs are starting to perform their nightly concerts of groans and peeps. All around me the smoke and air are warm. Nighttime during the spring, the summer (in the South) is different; almost magical.

Never had I ever imagined I'd end up in a place like Wilkesboro, North Carolina.

I did not think I would really end up anywhere - not anywhere worthy of something as strange as happiness.

Dogs are walking down the dark driveway, barking at their own footfalls and farts... a funny thing to watch, an even more interesting thing to be annoyed at, listening to them as their voices echo around me, me, up on the top of a mountain.

But annoying can be charming in its own unique way... there can never be a sense of home without annoyance. And anyone that tells you otherwise they are a liar.

Out of all the qualities in our home, my wife shines golden.

She will be home soon. I hope the flames welcome her with kisses of light and warmth. I hope my internal, bright flame will never be snuffed out, go cold and black and ugly.

Hermitage... A Wealth of Food... Connecting to the Living

Something about staying in my house for days on end appeals to me. I have been blessed with a life where I don't have to go out in public if I don't want to. I write from home and attend classes online. I bring in a small income from selling on eBay.com, two of my published books are still being purchased on Amazon.com, and my wife works full-time from home. Thankfully, my house has so much of everything I ever wanted; going out seems absurd.

Last night, however, Tara and I went shopping for this week's upcoming Thanksgiving feast. Talk about a shopping nightmare-spending-spree. Our shopping cart was bulging with vegetables, cleaning supplies, spices, fruits and meats, dinner rolls, candles, drinks - the works. When we got home and unloaded and put everything away, I looked at the fully stocked refrigerator and pantry, feeling rich.

"I think both of us have anxiety issues when it comes to food," my wife said.

Basically, she was reading my mind.

We have enough food to last a good while before having to go back out into the public. Having to go through the experience of *not* having food, *not* having the means to purchase food, and not knowing when your next meal will be is going to scar you. It has scarred us. Not eating for five or six days is a terrible experience. I don't recommend it to anyone. Now, I don't ever have to worry about that again.

Staying home - my safe zone - is becoming more and more engaging than ever. Yet, I will admit, there are days when I make a lousy hermit. Lunch dates after running errands, or simply because we don't want to cook, are still attractive and special to me. At one time, a time not that long ago, months and months (maybe a year) went by when I did not eat in a restaurant. That time changed something in me. Eating out is an occasion I very much value and appreciate... something that is not taken for granted. This past year, I was given the opportunity - the gift - of taking my wife, sister, and niece to a nice restaurant, and it was exquisite, special, and a fond memory I cherish. There was something about telling my wonderful niece she could do whatever she wanted on the menu that made my heart happy.

I still feel I have a lot of making up to do. Too much time was spent in my self-imposed exile. So, anytime I see family or close friends, I do my best to not screw anything up... and to leave a positive, memorable mark on the time spent with me. Sometimes, though, anxiety attacks happen at the most inopportune times (not that *any* time is a *fun* time to have an anxiety attack), and I feel embarrassed my mind and body have that reaction.

The famous French author and philosopher Jean-Paul Sartre once wrote: "Hell is other people." Believing that bullshit for as long as I had did not have a positive outcome on my life. I do understand what the writer was conveying, and, in general, I avoid the mass, unknown public. But there is another quote that bubbles in my mind, which, comically, is *not* from any famous scholar. It's from a character in a television series who was speaking to a very damaged and conflicted man: "You are looking for a way to make it hurt a little less... the only thing that helps is other people. Try connecting with the living." It all depends on what type of person one chooses to spend their time with, to share their life with.

Currently, every day, I am connected with a sublime woman I am honored to have as a wife. As difficult as I can sometimes be, she is there for me, believing in and managing me. I have no problem staying at home with her. Famous people such as Howard Hues, Axl Rose, and J. D. Salinger intrigue me with their reclusive ways. Maybe as I get older and even *more* set in my ways, I'll go out less and less. Before that happens though, I must, must, must visit Key West... if ever I was invited to Owl Farm in Woody Creek, CO, I would not refuse. Yes. There are some things I want to do and places I want to visit. On an average basis, I think I'll stick to my home, The Locke Nest Cabin, with Tara, the dogs, and the different fish in the aquariums.

Saved By Winnie, the Anatolian

There was a boy covered in a black sheet. His head was underneath. *He must be afraid*, I thought. I need to get to him, pull that damn sheet off his head. How can he breathe, for God's sake! I tried running to him. That didn't work. For some reason, my leg would not work. My right leg. It was like a weight. A weight with a chain wrapped around my ankle and an anvil attached to it.

I could not get to the boy.

I was trying and trying and trying.

But my LEG WOULD NOT WORK!

When I woke up, a heavy, hairy, warm, and familiar-smelling beast was upon me.

Winnie-Winnie-Winnie-Winnie had her big, beautiful, kind head on top of mine. A heavy paw was on my chest. And she slowly licked me awake.

I grumbled about something unintelligible and gruff – my throat feeling full of gravel. I looked at my watch: 5:16 am.

Holy good gracious, I thought, *I have slept in*.

Usually, four in the morning is when Winnie-Winnie-Winnie-Winnie and I have our "secret" time. The time when all the other dogs are asleep, and we raid the fridge and pantry for snacks and leftovers - FOOD!

"Sorry, girl," I muttered as my hand rubbed the top of her head and scratched behind her ear. "Everything is alright."

She continued to lay on top of me, trying to rub that mug of Her's into my chest.

The images from the nightmare stayed with me as I worked my hands through the big pup's fur, massaging her muscles.

She took her paw and started scratching at my face...

"I'm up. I'm up. I'm up," I whispered to her.

My wife was gloriously asleep beside me. I could make out her beautiful frame through the darkness. And I listened to her breathing in her deep sleep of peace.

I could smell her face cream, her shampoo - her body.

I was safe. At home. In Our bed. With a damn 80 lbs. puppy on my chest!

I did not want to wake her up.

How was I breathing when I couldn't save the boy from the sheet?

That was a question that already had an answer.

Winnie-Winnie-Winnie-Winnie woke me up from it - SAVED me from it.

We, the pain-in-the-ass puppy, and I walked down the stairs to the kitchen. She knew what was up. I was going for FOOD! FOOD! Glorious FOOD! I knew what we were going to have. I had grilled some sweet Italian sausages the night before. Complete with onions and peppers, soaked in a spicy-sweet sauce.

That's the ticket.

Winnie-Winnie-Winnie-Winnie seemed to have read my mind. She properly walked to the front of the fridge, sat her furry ass down, and waited, her tail thumping a bass line to a beat I was walking to.

We were going to eat. Our morning would turn into a fine day. A fine day, indeed.

For the First Time

There is a subject I have avoided writing about my son. That's right! I'm a father. Never had I planned on having a child. For the first three years of his life, I was very active with him, his primary caregiver.

Things in life change, though. Different things happen. People change. Situations change. There was a bad run I trekked for too long, a long road into and out of Hell, which led me away from my son for a long time.

For reasons which seemed noble at the time, they may have turned around and bit me on the ass. But it has not like I am unfamiliar with a painful ass nip.

I was not going to write about my son. That is what I thought was so noble. To protect his privacy, keep my life separate from his, and hope for the best, never considering the possibility I would one day *want* to write about him. Not because I was trying to "hide" the fact I am a father; it had more to do with privacy - HIS privacy. Exploitation leaves a bad taste in my mouth. And, I thought, if I wrote about him, it would violate some unmentioned vow.

Something has changed.

I have changed. I have changed a great deal these past two years, and so have others from my former, pre-North Carolina life.

It feels safe now to share with you, dear Reader, some more of my life. And that IS THE POINT of these silly little columns - to share and to help... maybe get a little help back from y'all. Now, that would be great because I have no idea what the Hell I am doing. I don't even pretend to know what I am doing anymore.

Fatherhood – shared parenting – living states and states away from my child… *nope!*

I have no idea what I am doing.

Last week, for the first time in over a year, I saw my son, Vincent. Tara and I packed up the VW Tiguan and drove hours on I-77 North straight to a hotel in Ohio for the special visit.

The hotel was to be the Kent State University Hotel and Conference Center, located in the heart of the Ohio college campus. What was important to me was wherever I got to see my son, it had to be *nice*. And *safe*. And, of course, have a pool and a hot tub.

I wrote about this experience in a Reporter's spiral-bound notebook. I was sitting in the hotel lobby, head down, and the pen flew across the pages that were being flicked over so fast it sounded like someone was dropping or throwing newspapers across the room.

(It was my nerves getting the best of me. By nature, I can shut my mind off from my surrounding environment when I feel I must and transport what I am thinking to a page. To be faithful and focused on that page… that is how I started writing in the first place.)

The gonzo piece was poetic, flowing with fancy language. It is a style of writing I have not constructed in a long time, but it may be too fancy for this format. I have never tried to be impressive, as in some stylistic, profound, or important writer, when typing these stupid little stories up.

What I know about them: they are good! They are good for me. Having to spend a little bit of time every day considering my life and having consideration for those involved in my life has radically altered my day-to-day life.

I am, at this very moment, writing this to you, dear Reader, acting with those principles that I just described. I *must*, for him, of all people, show consideration.

Vincent walked through those lobby doors, taller than I had ever seen him. And, undeniably, my son.

Tara, Vincent, and I had two good visits staying at that hotel.

I'll keep the details private.

The outcome, though, of this first visitation experimentation: I get to do it again.

After I bid adieu to all of you, I need to finish packing.

Tara and I get to see him this weekend.

Nothing like a Good Bookstore

It was at the desperate end of my final semester at the University of Akron when I discovered Last Exit Books.

Located at 124 E. Main Street #1, Last Exit Books and Coffeehouse, back when I first entered its door, was a small, quaint, messy, cluttered used bookstore that had that particular smell that bookstores and libraries have—a smell of leather and dust and wisdom.

What welcomed me when I walked through that door was a treasure. Hardbound, affordable books by authors I love and admire, not to mention learn from - with DUSTJACKETS!

My arms were loaded with titles ranging from Norman Mailer's Harlot's Ghost and The Gospel According to the Son to Ernest Hemingway's first edition of The Garden of Eden. There was a book of poems by Bukowski and other titles I can't remember. I think Augusten Burrough's memoir, Dry, was there. In fact, I believe that was my introduction to his work. That first visit cemented, in my mind, a vow... a sure marital faithfulness to this little bookshop of whenever I am in Kent, I will walk through the door, buy some books, and chat up the owner. And, for the life of me, I can never remember the wonderful man's name.

What is important is that I remember him, he remembers me, and I always go back. No matter what shape my life has been in, no matter how little or how much money I have, I make sure I go to Last Exit Books. I have been to hundreds of bookstores—some good, and some were so terrible and dirty and unorganized that it made me want to pull my own eyeballs out of my skull. And my mind felt like it was going to implode with all the lashes and zaps of knowledge of where and how the shelves should be arranged. A bookshelf - a specific shelf - to me, is like a canvas in which an artist has to take their hand and place the right line - the right title - in a specific spot, the next title following by subject, or year published, or where that title fits into a particular author's list of accomplishments.

What impressed me the most about Last Exit was the inventory of titles, the "at home" vibe of the place, and, whenever I'm there, it feels like I am the only one in the store.

Over the years, the store has tripled in size. There is a coffee shop as you walk into the store. The coffee shop alone is the size of what the original store used to be. Yet, there is still that good feeling I have inside me of the store and having our privacy. I'm very proud the store has grown. I'm happy it - an independent bookstore - is still up and running and expanding. Fifteen years (maybe more), my boots find their way back, walking through the aisles, eyes wide open and searching for the next treasure. And those wonderful first editions, hardbacks, and trade paperback wonders ARE treasures. To me, at least.

On Editing

I'm horrible at editing my own work—or anyone else's work, for that matter. I'm horrible at spelling, my grammar is atrocious, and I don't take the time to read over my work before I publish it.

I'm always so excited, listening to all the pages sliding and screaming into the printer's tray and the first time holding them (even clutching them to me, feeling that weight...the New Born is ALIVE – something I created), and the melting of something inside me begins. A once-swarmed desk... stacks and stacks of pages from the crime scene of my office has been organized and put away! All the reference material put back on shelves, loose-leaf, print-out pages filed, or thrown away; post-It squares – hundreds and hundreds of post-It squares bunched up in my fist, and, me, walking those fist-full wads of yellow paper out to the firepit feels a little free… a little sad, but anxious to get my hands on the new book.

Starting the next book – now, that is what must come next. And then - and then – more editing will need to be done. More expansions, cuts, and choices will need to be made for another Gift from the Gods.

As far as this First Draft of what will be my third published book, I have to keep working on the changes, expansions, scenes needing to be reworked, and stories completely deleted. Tara and I tell ourselves how old we are becoming. Or, more accurately, how old we are. I don't just feel like I am getting older. I can tell by the pace at which I am producing The Word.

Slowing down has been the hardest thing for me when welcoming myself into the Locke Nest Cabin. And I do not mean physically slowing down; however, that is a part of it.

I love getting the words down and getting them down fast. That is what I live for. The moment the thought, images - and words - come to mind, when they travel through my system to the tips of my fingers, and - BOOM! There they are! On the page!

Thinking the way I do, which is scattered (at best), moving my fingers over the keyboard... It had to be done fast, all the while always in the hope of getting that buzz and falling into a private world- a world that manifests itself right in front of me.

All you writers out there know what I'm talking about. Trying to maintain that: making that "high" last as long as possible. What can I say? I have an addictive personality disorder.

Words are no exception.

Why am I this way? Why do I treat something so dear to me with such irresponsibility?

The simple answer: because I can. Because it is who and what I am.

I love driving down the page in reckless thought - in both violent and considering, thoughtful ways. The First Draft, no matter what anyone says, is the most important - the most REAL! It's a raw expression, like a person, complete with flaws.

It should be really important when (as we all know it is) over the gift. The Great Moment came—time for cosmetics. The foundation has been built.

In the past year, I have had to look at over five manuscripts I created (some still need to be finished) - manuscripts that had been read and edited by others - and I had to change/improve and, in some cases, destroy the original. And that is not a BAD thing. Hell, work, like all things in life, evolves and becomes better through change and improvement.

With me, though, on a regular day, I want the words down: type, breathe, spit, light another cigarette, and move on to something else. I am one impatient person, and some of the elements of my life are very... energetic... even though I'm an echo in the shadows of how I worked years ago.

For example, last night, I was writing a column about building a model motel from Alfred Hitchcock's 1960 classic, *Psycho*.

The beginning of the piece was somewhat good, with very few flaws in punctuation and spelling, nor were there any spaces a word should have occupied. Then, as the fifth paragraph morphed into a sixth, a seventh, an eighth, the MAJOR flaws started sprouting like embarrassing blackheads on the prose. It's always the next day, the next night after someone else has read the work, that those ugly blackheads are made visible to me. When they are pointed out to me, and, yes, I am grateful to my free-speaking, opinionated editors... without them, I would not be the writer I am today. However, the embarrassing and the terrible and downright humiliating thing about all of it is I KNOW BETTER!

That's right.

I know I should reread the work, edit it, expand on it, grow it, and mature it.

All manuscripts must go through various revisions. Reluctantly, with a dash of bull-headed stubbornness, I will make those changes, those improvements.

Why don't I do it with more vigor and spice and zeal?

Maybe it's simple laziness. Maybe it is excitement.

(I think a little bit of both. Yes. That's it! Too lazy to fix up the Junker and too eager to drive through time into the next trip.)

What I do know is I love what I do. Believe it or not, it is hard work with little to no pay. The major reward is FINISHING a piece. When that last period is punched on the keyboard, and the sigh of relief knowing you have done something and done it well... well, hell, that beats cashing in a check. For me, it does, anyway.

There will always be mistakes. Learn to live with them. Do not just do what I have done too many times in my life: learning to live with mistakes is important. Learning from mistakes, your own, as well as others, is called an education—an education through experience. Your mistakes do not define you. They simply show forth the human aspect of you and others.

And TRYING TO CREATE SOMETHING OF YOURSELF IS NEVER A MISTAKE.

There Will Be Blood

She is not a nurse. She is a phlebotomist. I'm not sure what that is, let alone how to spell it. (That's right, I had to use spellcheck on that word and read the definition in the dictionary. Simple definition: a vampire that does not DRINK blood; they just acquire and collect it from willing and unwilling victims.)

That's right. This woman was here to take my blood.

I'm sick. I'm in a Baptist hospital in Wilkesboro, NC. Blood testing is a must for my mysterious condition. Severe pain on two sides of my lower back and sternum, pain that doubled me over, made me moan terrible, bullfrog-like sounds from my mouth, and caused the *fear* of knowing something awful was happening. Something I cannot control – AND IT HURTS! All of that led me to the second floor, room 217, at the Baptist hospital.

My blood will tell the doctors more about my ailment, I hope because I need some answers. I'd already had X-rays and screenings. Now, it is time for the blood to be drained.

Her name is odd. Reading her nametag, I vowed not to try to vocalize it. I was having trouble concentrating, anyway - no need to embarrass myself by sounding like an illiterate unable to verbalize a name.

She saw me starring at the tag:

"It's French," she said, pronouncing her name for me.

She then laughed, looking at the contents surrounding me on the bed and bedside tray: pens, stacks of manuscript pages, and two overturned books, along with my laptop.

"Are you a student?"

Her attention never really left my arm. She may have been aware of the details and items surrounding her, but her focus was dead-center on a vein. Her concentration was terrifying.

"No, no. I'm a writer," I say as she works the needle beneath my skin. The pinch and sting from the needle made me breathe in deeply and pushed my eyelids to fall like a theatre curtain, taking away my vision and the image of my arm being pierced for the seventh time that day.

I'm a "hard stick," as they say in the medical community. It takes various pokes to get at one of my veins. Just to *find* one of my veins is a chore.

"My great-grandfather was a writer."

Nice.

"My father looks just like him. When I'm done with this, we will look at their pictures. I have a few on my phone."

I'm game.

She went on to tell me he was a "big deal" in his time. He influenced many famous writers of his generation. One of them, poet Charles Baudelaire, was a big influence on ME when I was in college, and going all the way up to my early 30s, Baudelaire's life and work fascinated me.

After she found the vein, three tubes were filled up. I had to exhale a weight of thankfulness from my lungs, happy the ordeal was over.

Afterward, she was on her phone. We looked at photos of her father. She then pulled up this:

Pierre Jules Théophile Gautier, PHOTO by Nadar

Her father can indeed pass as Mr. Gautier's twin.

After she had been gone awhile, I spent some time on the trusty iPhone, researching said great-grandfather. I also ordered Mr Gautier's *Mademoiselle de Maupin*, his 1835-36 novel, which, apparently, is a shocking tale of sexual deception that "draws readers into the bedrooms and boudoirs of a French chateau in a compelling exploration of desire and sexual intrigue."

The novel was cleared of obscenity in 1922.

Good. Good, I thought as I placed my order. I love reading banned books or books that *were* going to be banned.

Education, in my mind, is the ability to acquire knowledge and insight through exposure and time. My education on Pierre Jules Théophile Gautier has just begun. Knowing me, in a year's time, there will be a section on one of my bookshelves dedicated to his works.

It's a funny thing: having a famous French author's relative withdrawing my blood, introducing me to him and his work and life. My life seems to take many odd pitstops along the way to wherever it is I am going.

To the Gautier family, thank you for your roadside attraction along my journey.

Dear Great Pumpkin

October 18, 2022

Locke Nest Cabin

To The Great Pumpkin:

This is the first time I have ever written to you. Growing up, watching the cartoon, and feeling bad for poor young Linus and Lucy waiting for you to show up (which you didn't) did not exactly put forth any confidence in me you would read anything I sent you. Or show up on Halloween to deliver any gifts to me.

But now, I'm in a dire way, needing a gift from something or someone great.

This year, I'm going to send this long overdue letter. I'm not writing this to get gifts rained down upon me in a pumpkin patch. Not loads of books, or weapons, or anything else I collect and cherish. Nor am I writing for an explanation as to why you kept leaving poor Linus hanging on to his devoted faith in you without showing up to his most sincere pumpkin patch, to his most devoted self in the belief in you.

(Then again, maybe THAT WAS YOUR POINT! Sometimes, all there is to go on is faith and faith alone. A lesson in endurance and unquestionable belief. Maybe that is the gift you showered upon him: a lesson in faith, and faith alone.)

So, I am not writing for presents. I'm not writing to get a response. I'm writing to thank you. Thank you for all the years I have been watching and waiting for you. Right there, alongside Linus, I watched. All year long, I waited for Halloween with an eagerness that matched Christmas. (Maybe even out-did Christmas.) And I waited to watch that silly cartoon, with YOU as the main attraction, and you always were a no-show.

Also, this letter is for all the children that believe in that fantastical kindness and never-dying hope in the young of an entity of great goodwill. That is a quality too quickly dispatched in all our lives.

Wouldn't you say so, Great Pumpkin?

For Halloween, just this once, rise out of some little pumpkin patch in the heart of a child. Deliver the gifts only joy can bring. Halloween is more than a simple tradition of dressing up and causing harmless havoc. Halloween is one of those nights when imagination is jump-started in people of all ages. It is also a night that can bring the memories - good memories - of childhood back through smells, through visuals, and through the passing-down of joys restricted to that magical night when witches fly through the sky, orange smiles glow in the night, skeletons dance their merry dance in graveyards, and laughter is passed through masked and painted faces. The night of dancing skeletons and shrieks is a joke and joyous on a moonlit Fall Halloween night.

Dear Great Pumpkin, thank you for all these years I have had with you. And I will pass the flaming jack-o-lantern grin of fun and fantasy along to any child needing some good old-fashioned October fun.

Till next year, with affection,

Leland

PS: The gift I would so desire and be grateful to receive. Please help me not go to the hospital anymore… I don't function well in any type of institution, especially hospitals, places determined to put hoses and needles into me. I know they are there to help me. I'd rather stay at home. Or wait in a pumpkin patch for you.

Just Like a Kid

Going out to get pizza is something I have enjoyed since childhood. Usually, I think of movies, Friday nights, and the start of a weekend... you know, kid thoughts. Thoughts of summer afternoons and after-school nights with friends at the pizza parlor, the Book It! reading program, and pitchers of Coca-Cola filled with ice. Smells of garlic and sauce and baking crusts mingled with the greasy smell of chicken wings being fried and the hot sauce sprinkled and coating them in plastic baskets lined with wax paper, garnished with celery, white ranch sauce, and carrots.

Tonight, my wife and I are going out to pick up some pizzas. No! It is not Friday - it's a Monday evening, in fact - and we are not kids. At least our bodies aren't. However, dear Readers, I would be lying to you if I wrote that I do not feel like a kid. Just this afternoon, I even received a toy in the mail.

(It's an adult toy, but a toy, nonetheless.)

Now, I get to eat pizza on a Monday night. Hell, yes!

Who said you had to grow up to have fun? And I don't have to wait until the weekend to start the festivities.

(I think I did, at one time, believe you; you had to grow up to have fun and to have that magical feeling of freedom… but I was wrong. As I usually am. Fun can be the re-living of days past and gone. Days that were simple yet complex in their very existence. A feeling brought back by smells and sounds and movements. And *tastes*. Biting down on a thick slice of good, hot cheese pizza with the proper tanging tomato sauce, or hot, soft pretzels, movie theatre popcorn… It is enough to take anyone back to a time that was good, hard, confusing, and memorable. A mind is something that can hold the small details of a much larger event.)

It's a small and silly thing: getting a pizza. On the other hand, if it's something that brings a smile to your face, makes you, for a moment, forget about the bills to pay, the projects to complete, or the struggle to keep it together, by all means - buy that cheese and greasy, crispy pie.

You will thank me when you do. And remember, not all great things are hard to obtain.

Great things stay with you. Even when you think there is nothing to put a smile on your face, remember silly movies, French fries at the fair – and hot pizzas from your local pizzeria.

What a Life

It was not that long ago when the cold weather was something I feared. And rain, as well as the blistering hot sun and humid air of an Ohio summer. I lived and slept outside. Elements of the environment around me were major factors in my comfort, mental health, and physical health, as well. Some days and nights, there would be people - strangers that would take me in for a night or two; sometimes, they would keep me for a week. Sometimes it was a good thing. Sometimes it wasn't. Most of the time, it wasn't. The perversity some minds harbor is amazing.

Food was something I did not think about. Nor was it something I really wanted. My mind was focused on putting something else in my body.

It was hard times. I lived hard and bad with hate and frustration in my heart. Those things – frustration and hate – were the two things I held onto the most. They were like the electrical shots that kept my heart pumping, my mind focused… focused on something, anyway, besides my addictions.

For all those who know those feelings, understanding when to put them away is essential to recovery. Another important fact is recognizing the many miracles and gifts a whole life has.

Something as small as a hot shower or a snack at night while watching a movie – CLEAN CLOTHES – all play a part in separating one life from another. A better life is obtained and defined not through possessions, wealth, or capitulated peace but through recognition of the miracles that seem to spring out from nowhere.

Is my life full?

It is full enough for me not to be fooled into thinking my heart, life, and memories should not be shared.

True Story: once upon an October evening, as a man is pondering and weary, before turning off the light of his library, he looks around his great and cherished room - his pride, his collection of books and memorabilia. The year before, the man owned three books... now, there is a library.

(His own collection of writings has been published, and THOSE books are now included in the multitude of volumes on his shelves.)

Jack o' lanterns smile there glowing, gapping smiles with fire for eyes; skeletons dance to a merry melody as winds chill the night air, and witches fly through the night...

What A Life!

A life still being lived during the Halloween season.

To be warm in the orange glow of my cabin, a safety felt in the calming lights.

I have stuck up my middle finger to REAL monsters that scream in the night.

I know they are still out there. Still selling all those poisons, turning many into ghosts and ghouls and shivering goblins wandering the wet and chilly midnight streets.

Let them have their nights in manmade hells of differing stages of greed, anger, and sloth.

I'll keep my corner of the world warm and safe. They – the real demons – can have their trash and misery. They deserve it. And I'd rather them stay there than be anywhere close to me and mine.

They will remain where I know they dwell. God help them if they come anywhere close to me or my family.

Cigarettes & Salem

My mailbox was full... that, I am sure. As for the rest of the night, it could have been madness - the kind of madness sinking and rotting into the soul when the soul has been spent and left in a city.

After days and days spent in a hospital, knowing no one had checked my mailbox for packages, I was sure the freaking box was full – filled with bills, junk mail, and packages from online shopping orders.

The city of Winston/Salem. That is where I found myself two nights ago. It was dark: past nine at night. Security Guards were checking permits, and doing their duties, wearing their white negro smiles with their batons ablaze at their sides.

And what was I doing?

I was looking to smoke a cigarette. I had "permission" to leave my gloomy hospital room, go outside, and smoke. You would think a city such as Winston/Salem would support such behavior. It is the American Tobacco Motherland. Hell, two brands of popular smokes originated in that concrete jungle of hypocrisy. Because of nicotine junkies like me, there is such a city, never mind the hospitals the cigarette companies paid for.

What started as a low - a low, low evening, in wait - turned even lower, still.

God bless the big man from the gas station (he gave me the book of free matches.) That was a long hike from the hospital. It was a BP gas station. He would not permit me to use a lighter simply. My pockets were already carrying around discarded, half-and-quarter-length cigarette butts from off the streets and in ashtrays outside of closed businesses.

This was a mirror to my past - and its desperation. Flashes of my history in Youngstown and nightmares from Cleveland were working their way through my head. Feelings and thoughts from that time could not be stopped from igniting a sinking feeling in my mind… memories I hoped never to be bothered with EVER were winning the battle.

I have walked MANY streets. Been down some dark alleys... the type with glowing eyes watching you, sizing you up, trying to determine what you might have, and razorblades clutched in fists, waiting in the corners and in between rotting, coffin-like buildings.

I'm not sure what came first. The dumb, stuttering, and stupid, pathetic attacks... or the thought of rationalizing them. What could not - would not - be rationalized was the horror with which my wife found herself. A man - her man - sick in the brain and still wanting nothing more than to fill his cup with overflowing addictions and predilections...

It was she who made sure I made it to a hospital to be worked on. And, oh, how she cares. She really does. The miracle of it is her love seems to come from a well that never goes dry.

My hundred yes, yes, yeses to the sky and to the air around me served as my mantra to go on. To finish this rotten endeavor, I'd put myself into smoking a damn cigarette within 30 minutes. Then I need to return to my hospital room, back to the mechanical bed and mind-draining television sitcoms and commercials and incomplete, edited movies. Edited and "corrected" four-letter words, images, and scenes of violence and nudity.

I was NOT looking forward to going back to the room, the endless edited television, and the many "check-ins" with the nurses throughout the night. The damn blood-pressure cuff, needles piercing me for blood...

Truth be told, I was LOST. I walked around and around the massive, massive medical building after smoking my trash cigarette butts.

(I think I entered into seven different ENTRANCES/EXITS.)

Just being lost - not in the city - just LOST around a building IN THE MIDDLE of the night was crazy enough. I was out of breath, hot, and feeling lonely... and all the wonders in between served me well. The sweating was terrible. Large drops of it fell and made a miniature river from my face. My hair was drenched and sticking to my skin.

I DID NOT have to be in a hospital.

That is what I came up with.

I NEED TO BE HOME!

The one place where SAFETY is not only felt but attained, and peace is something not merely discussed – it is felt, and it is REAL.

Dreaming About It Happens from Time to Time

She was in my dream again. I'm not sure how many times I have dreamt about her throughout these ten or so years since I first found out about her. She always seems to remind me I have a job to do and a gift/promise to her I told myself I would deliver. It was during a hellacious June month her story captivated and horrified me.

Her story being years and years older than my own existence, one would think I could feel far away from it, protected from its horror and general, disagreeable details, and I'd be able to sleep soundly, leaving her and her damn story in the past. But I can't do it. For the life of me, I just cannot.

Months and years have gone by when not so much as a single sentence has been created by me to share her story. In a drawer is a stack of disorganized, incomplete manuscript pages. The bulk of the work has been lost - lost during that time in my life when ALL seemed to be lost (and many things were lost) and held no meaning. However, I still had dreams of her, like I did tonight. I had not even been asleep three hours before my eyes were open, searching the darkness surrounding me for something moving around.

There was, as always, nothing—just the thoughts and memories of the dream.

After a moment, I was getting up from bed, leaving behind my bed full of sleeping dogs and a wonderful wife. I had to leave. I knew after a dream like that, after so many of them over the years, there would be no point trying to get back to sleep. Waking from a dream like that - a reminder such as that - sleep has a way of leaving me the way warmth can exit the body when the heavy, comforting covers are thrown off.

It's best just to accept it's going to be one of those nights. It was a night of me pulling up the saved files on my email account and rereading the facts of the case. And, with all the dedication left in me to continue with the project, I brainstormed some new brew in my brain on how to proceed.

I even keep a framed photo of her to the left of my laptop. It is a print-out image from a newspaper published in May of 1964. Her name, Ruth Gutherie, is in bold caps letters beneath her image.

How many hours have I looked at that damn photo? How many hours of pulling at what magic creative strings I may have left to try and conduct some semblance of her last days on earth have gone by? This is, without a doubt, the longest amount of time I have spent thinking about a story. And, in all honestly, it has to be more than just a story I'm writing. It's a memorial. That is how I feel about it. It is a memorial to a girl who, in all the harshness of how life and the world can be, could easily be forgotten. Her murder, like so many others, has become nothing more than collected data at a local library—an unsolved case in the files of a police department.

I'd like to change that.

So, dear Ruth, we are still in this together. I have not forgotten you. And your story will be known.

A Little Bit of a Different View

Years and years have passed. Initially, I was going to write about this subject - this film - back in 2004, when it was released to the world. And the world responded and responded. And kept the controversy of its very existence alive for so long, I wondered when it would finally go away. When the new, next-day headline would, as they ever do, push another impassioned topic onto the front pages, casting aside the old, making way for the latest to begin.

And, eventually, it happened.

Mel Gibson's 2004 The Passion of The Christ was, without a doubt, one of the biggest films to come out that year. According to an online source: "The film was controversial and received polarized reviews, with some critics calling the film a religious and holy experience, while others found it anti-Semitic and criticized the violence as extreme. The film grossed over $612 million worldwide and became the fifth highest-grossing film of 2004 internationally at the end of its theatrical run. It is the highest-grossing (inflation unadjusted) Christian film and independent film of all time."

At the time, all I could think about was the controversy of the film. How much passion and anger did the film instill in the hearts and minds of many...? True: all good writing and art have a particular responsibility to shock, to make people think, to make them FEEL! However, after my first viewing of the film, I found it to be distasteful. Plain and straightforward: distasteful. By the time I was 20 years of age, I had seen enough gore and horror films - all of them a part of some separate category ranging from ridiculously absurd to borderline masterpieces of cinematic history - and this film, in my twenty-year-old mind, was no different. I thought Mr. Gibson was perversely cashing out on a story that means so much to so many people in various ways that, to me, seemed pathetic. That is genuinely what I thought and how I felt.

Now, tonight, a few months away from my 39th birthday, having seen this difficult-to-watch, 2 hours and 6 minutes film for the second time, I recognize a change in me regarding Mel Gibson's adaptation and interpretation.

How many have been brave enough - or arrogant enough - to bring to light their own vision of the final hours of the Life of Christ? I know I would never try to take such a feat on. For starters, my own devotion, faith, and personal philosophies tend to hinder me from completely swallowing such fantastical concepts - mainly the one being a singular man, a carpenter from Nazareth, who died in such a horrific way for my sins- for the sins of all people. But it is not Gibson's bravery or arrogance that pulled at the different strings in my mind tonight regarding The Passion of The Christ, *n*or is the copious amount of violence in the film one of the issues I struggled with during this second viewing.

It was having to mentally put myself in the boots of the actors, director, writers, and so on to recognize this film was, in their own way (maybe) their passion. It was their love letter to their Lord. It is a gift for the tens of millions of followers of Christ to have a visual, comprehensive product/tool/reference to what may be viewed by so many as the ultimate sacrifice ever recorded or written in Western Civilization. It is not only an accomplishment, but I can also now see it clearly for what it may be: confirmation of what they have put their faith in.

Something becomes more real, I imagine, for some, if it is a movie. And not just any movie - a movie made with dedication and backing and love and, dare I say it, passion.

To expound on my own personal views, I believe Jesus of Nazareth was a good and holy man. A philosopher and controversial thinker at a time that needed such a person. His teachings, undeniably, are one of the most important vertebrae of the spine of history. He is the one who will be remembered and followed probably for as long as the human consciousness continues from one millennium to the next.

At least, I hope it does. And I don't know why I feel that way. I am, by no means, a preacher to the ways of a pristine Christian lifestyle and following. To the contrary, truth be told. I question way too much - not in the goodness of The Good Lord - but in people and their sometimes blind, unquestioned following. Some things that, initially, were meant to be a thing of absolute beauty and hope have been corrupted, turned obscene, and tragically disfigured.

Man can screw anything and everything up. Think about it: Man turned water into an explosive. The hydrogen bomb, for goodness sake, is proof enough that the most precious, natural things on this earth can be twisted.

Water, ladies and gentlemen is something every person, no matter their faith (or lack thereof), needs. It is an essential element to survival. Faith and belief, in all honesty, are things not required to survive physically. Belief in whatever it is a person believes in may be the most important thing in someone's life, but just remember, it can be skewed and made toxic by the wrong people.

Getting Ahead of the Game... A Different Kind of Year

It goes without saying my wife and I are not traditional people. Our values are common, if not distorted in design, by our own personal conceptions of what values should be. We live and work in a non-traditional way, staying at home, fawning over and discussing our large family/pack of dogs, and having long conversations about how our pasts have formed (or deformed) how we feel and act and react. Tara seeks solace in watching the fish in our large aquarium. She is shamelessly kind and cares a great deal about not hurting anyone's feelings... avoiding conflict and confrontations is an art form she has mastered. However, she can be as fearless as a one-hundred-year-old snapping turtle striking at a bulldozer.

And for some reason, probably something I will never understand, she fell in love with me at the age of 13. Twenty-some years later, she tracked me down, drove through several states, picked me up from a horrid and dangerous section of Cleveland, OH, and claimed me.

She more than just claimed me.

The next day – the day we crossed over into North Carolina - she filed for divorce from her then-cheating husband.

We barely have left one another's side since. Nor do we ever plan on being with anyone else. It's just not an option. And we are of an age where we know who we are and what we want - the developmental stages of our life, along with our difficulties and peccadillos, were cemented long before that dreary-grey-yet-miraculous morning last November when she got me.

Now, for a woman who is non-traditional in many facets of her life, this year, she wants a TRADITIONAL Christmas!

She has already purchased an 8-foot tree, a thousand twinkle lights and ornaments, and hundreds of presents, and from the light in her eyes, I can tell she is very, very excited about all this.

So am I, for that matter.

We had a wonderful holiday last year. Holidays, to be exact. Thanksgiving was a little awkward; I was just then being introduced to Tara's world and the people that inhabited it.

Christmas, on the other hand, was very special. It was my first REAL Christmas in years. We had a small tree, gifts - the works. We decorated the tree; classic Christmas cartoons were playing in the background.

This year, something is different. We have our new home, The Locke Nest Cabin, and, I imagine, there is no reason not to really deck these darn halls with joyous, psychedelic decorations, litter the floor beneath the tree - beyond the tree - with gifts... later to be litter - or should I say FUEL for a fire - cook a feast for visitors during the mid-afternoon/evening hours, and make calls to out-of-town family members, asking them how their holiday is going.

It's not even Thanksgiving yet, and Tara has all of the shopping done. She is determined to have the tree up the Wednesday after Turkey Day! Madness, I tell you, sheer and utter, blissful and magical madness.

Who am I to put a damper on such a wonderful, classic celebration?

There is also the generous, caring side of the season uncalled for, which I know we will perform. The dropping of bills and coins into the buckets of bell-ringing Santas and volunteers outside of stores and malls and Main Streets. Going to hospitals and/or churches to donate our time, care, and efforts to the community... to pray for forgiveness and thankfulness during the final weeks of the year...

Halloween may be my favorite holiday, but I will not deny the otherworldly, warm, and distant satisfaction I feel during December. A sensation that comes only once a year.

This second Christmas with my girl is going to be uber-special.

It's the first year we get to celebrate it in our - and I do mean OUR - home. The dogs and puppies and fishes, I bet, will be wandering around the place, wondering, *what the hell is going on?*

I imagine Winnie, Winnie, Winnie, Winnie, Winnie, our still-true puppy, will chew on wrapped gifts, steal tree decorations, or rip the stocking off the wall, thinking they are her special toys.

I can't wait for the chaos. Also, I can't wait for that long sleep on Christmas night, knowing things are the way they are supposed to be.

(Author's Note: Winnie-Winnie-Winnie DID tear the place up, chewed up gifts, stole decorations, and I think she shocked herself chewing on an electrical cord.)

A Quandary

The name is not as simple a thing as people assume it to be. Especially if you have so many. There is the fundamental rhetorical question - the Shakespearean question: *What is in a name?* There is your given – "Cristian" name – bequeathed your birth. Then, there is the common shortening of the name to a nickname. Nicknames come from friends throughout the years, some as a title, and others as a simple description of one's personality.

I have had many names.

Andrew

Andy

England

Drew

Tex

Leland

Lee

Lee-Lee

Mr. L

Eli

... I'm sure there are more...

It wasn't until yesterday that I realized how important a name is. - More importantly, how much it can mean to another person... how it can make them feel.

I have a sister. A wonderful sister I wouldn't trade in for all the answered prayers throughout history. Let alone the unanswered ones. Now, normally, since I was twenty years old, I introduced myself as Leland. That is my name. My middle name. But LELAND was also a character; it was a mask to wear, a manifestation of misbehavior I could manipulate.

(Don't ever call me LELAND in front of my sister. Those little annoying and uncontrollable hairs on the back of her neck stand up in fury and frustration. And knowing my sister, she wants to correct the misunderstanding. I fear one day she will correct someone at the wrong time, in the right place, under the rightful knowledge of who her brother is, what his name is... and, the funny thing about it, she will be right. And I won't be able to say a single thing that will deter her from expressing what is right and what is not so right.)

To her, to my family, I'm ANDY. Or the horrible and doomed DREW. DREW is not a character. It's a vocalized prison inmate number put into a title. When my mother calls out DREW in one of her sinister, yet-oh-so-silent-way, I'm in trouble. I'd screwed up in some embarrassing, juvenile way, and I better pay attention to what she had to say; otherwise, I'll be digging myself into a deeper grave... lighting the match to ignite my own funeral pyre, and so forth.

I'm almost 40.

Hearing that haunting, single-syllable word – DREW! - still, to this very day, straightens my spinal cord, flatlines my fury, and forces me to pay attention.

Shakespeare may have had a point. A good point. A word - a name - can be irrelevant. "What's in a name? That which we call a rose by any other name would smell just as sweet." That may be true. But Shakespeare never met my sister. Or my mother.

My name is Andrew Leland Locke.

If ever I offer my hand for a shake, I will always be thinking about what name I should give... what person I should be. It should be an easy question to answer. Just be me. Right? Well, the bitch of it is ALL those names and personalities are ME. Parts of me rearranged, put together, taken apart, lashed back together, and presented.

I guess that is why I like to stay home so much.

When I asked my wife what my name was, she said: "My husband. Mine."

I can live with that. And that is the name I want.

And So It Goes

I did not wake up angry.

When I looked into the mirror - when I looked into my old eyes - there was no anger.

Sure. I did not sleep well. I did not sleep very long. A few hours, at most. I woke up thirsty. A dry-mouth sensation - an irritation - was there. It was the image of the machine and the feel of the keys needing to be punched that got me up. The typewriter was waiting for me, but I was not annoyed. I looked forward to the work. It was dark/night, early morning.

Everyone in the home was asleep. It was nice, walking down the steps, the warmth of the living room welcoming me, my spot on the couch open and promising...

My typewriter was on the side table, to the left of my special, comfortable spot.

I turned on the television, chose a movie I would not pay attention to, and put the typer on my lap. And I knew - I knew something magical was about to happen. The music was forming and melting in my mind... something was happening. That feeling of knowing something was about to go well is so intoxicating... the story (whatever it may be) is there, waiting to be told and typed up in fury and passion.

And it began.

The words shot out of my fingers. Sentences were slicing their way into being.

Music was playing somewhere in the background. Explosions and dialogue were going on somewhere in the other world... in the other world, dogs were wandering around her property; she was upstairs, and my mind was on the Word and the words sprouting up in front of me.

Then it was over.

A dreadful scolding pulled me from the work.

Something about an alarm, something about terrible music, about sleep, and terrible days of no sleep - not enough sleep - shattered the contract I had with my work...... work!

It no longer feels that way. It's as though the time I spend jerking off and around and pounding away on the keyboards - with the pen - is an afterthought of doing nothing.

It can be interrupted.

It may not be essential, like sleeping, eating, and drinking. It's what I do, though.

Oliver Tramp & Trash

Sleeping next to five dogs does have its benefits. It also has its interesting, uncommon moments of intensity.

For instance, early this morning - around four - I woke up to Buck the Beagle sitting up at the foot of the bed. Usually, almost every night, he is asleep by my head or curled up by my left side.

To see him sitting up, alert, and making strange, whining, and growling noises put me on edge - it woke me up in a state of immediacy... for what? I did not know. But something was going on. And I had to be awake for it.

Something was a shift out of place or not belonging.

After getting out of bed and turning on the living room light, I went for my Smith & Wesson .357. Buck stayed in front of me, panting heavily; large, raindrop-like puddles of beagle saliva fell from his jowls, and I could tell he was on edge. We walked through the living room, stepping over five much bigger dogs, blissfully ignorant of *anything* that was going on… and we moved to the kitchen.

I turned the lights on.

Nothing.

An exhausted refrigerator motor was running and humming. The sink was dripping. A stack of mail sat on our long dining table. Nothing seemed out of the ordinary, yet Buck stayed in front of me.

I switched on another light, leading down the hall, ending at the side door of our cabin.

Buck growled in a low and violent and threatening challenge to something I could not see nor had any inclination of.

I cocked the .357. I moved closer and closer to the door.

What was I going to say when I opened the door?

Was I about to shoot someone?

How would I explain the situation to the police? A lawyer? A judge?

No need to concern myself with details, not at that moment. My job was simple: find the problem and make it go away.

With all the courage I could muster, I turned the knob, pushed the door - more like threw the door - violently open, aimed the pistol...

To nothing...

Nothing was standing in front of me.

No bad men. No monster. Nothing to shoot to save me or my family.

Then I looked down.

There was a teenage raccoon looking at me. A ridiculous young raccoon with bright, black eyes, innocently looking up at me. I didn't even point the revolver at him.

"Shit," I said.

The little thing didn't move. Just stared at me.

"What the Hell, man?"

He did not respond. He just looked at me with an expression of surprise. Just as surprised as I was. Nothing happened. We just looked at each other.

I turned around, shut the door, walked down the hall, turned the lights off, and headed back upstairs to my bed.

The next morning, I told my wife about the encounter.

"Oh my God! I'm going to make a bed for him. What should we call him? I need to put some food out, too."

I could not believe what I was hearing.

And that's what has happened. There is now a WELCOME mat on the side porch for any - *and I do mean any* - nocturnal creature roaming around the North Carolina mountains at night.

Tara has named him Oliver, as in Oliver Twist, the classic abandoned child of literature.

(I approve of the name, at least.)

I just hope the dogs don't get him.

A Year Later

December 24, 2022 - Wilkesboro, NC

After living and moving in three different locations in the past year - three different houses and a trailer, even a camper - I'm finally able to sit down to write a Christmas Eve column in my home... hopefully, my permanent home.

Its 3 degrees outside in North Carolina's Appalachian Mountains. The weather channel says it feels like -7. It's very warm in the Locke Nest Cabin. It's warm enough to make me sweat as I'm typing this. The dogs are panting; Tara, my wife, is sleeping and smoldering away in a bed warm enough to bake bread and biscuits in.

Our plan for today is to stay home, love on one another, "tolerate" the dogs, and open up one - just one! - present, and bask in the blessings of our life.

A year ago today, I was doing exactly what I am doing right now at roughly the same time. And, same as last year, Tara is safe, comfortable, and happy living with me. In the past year, we have gotten married, adopted two more puppies, and have the start of what I hope to be a great wealth of wonderful memories of a wonderful life.

From what I am told, my family in Ohio is doing well. And that's enough for me to know.

During the Christmas Season, in my family, I know certain things - traditions - will happen. If they didn't, I would really, really worry, like a person who has rationed out their final purchase of groceries, knowing one day the food will be gone. I know every year my mother will bake the best cookies. Several kinds of cookies she has perfected over her many years of baking them. Practice does not make perfect. No. No. Practice makes better. Experience gives birth to perfection. Unlike sheer practice, experience adds that extra little kick of knowledge only attainable through a much more mature amount of time.

Trust me, my mom has perfected her Christmas cookie-baking gathering with family and friends. That is one of the more beautiful qualities of her creating those haystacks, baking the clothespin and sugar cookies, and rolling up the chocolate and peanut butter ball buckeyes.

I know my sister and father will go on their yearly Christmas shopping date, buying gifts for my mother to open on Christmas morning. It is a tradition that has always made me happy and smile to think about. Hell, it makes me chuckle out loud as I'm typing this, trying to imagine what those two are like together in the store aisles, pushing a cart in front of them, discussing back and forth to one another about whatever it is my mother wrote down on her Christmas list to her Santas.

As for me, I started my traditions - both new and old - a year ago. I started watching my holiday cartoons (*Garfield's Christmas* Special, How the Grinch Stole Christmas, and the like), decorating a tree and a home, sending out gifts, giving to others I hardly know, and having phone conversations with those close to me. That warm and safe Christmas feeling, ever as intoxicating as anything I ever ingested, has been gifted back to me.

And I thank Tara, that amazing wife of mine, as well as my best friend, my parents, my sister, and her husband, Shad; those two breathtaking nieces of mine that have not a clue how much they mean to me and Tara. I thank the newest members of my family: Shane, Dawn, and Alex.

And I thank God for it. For *all* of it!

When I Need A Break From Writing

There have been many times in my life when I wished my father had kept night hours. Many people have been bothered by my middle-of-the-night phone calls... some of those calls are simple, "Hey, how are you doing?" interactions. Then, there are the other types of calls that have a specific purpose. An actual question needs to be answered, or I very much want some support.

Tonight, this is one of the nights I'd like to call my father.

Nothing terrible is going on. I'm not hurt or sick or in trouble. My wife, Tara, is fine and sleeping peacefully upstairs. My bills are paid up for the month; Christmas was wonderful - everything is okay.

However, I'm in the middle of a modeling project. I bought a wooden HO-scale Bates Motel model kit. (Earlier in the year, I put together the Bates mansion and wanted to have the motel, as well. My goal is to assemble a diorama, complete with landscapes, stairs leading from the mansion to the hotel, a parking lot, trees, grass, and, hopefully, some LED lighting.) The problem is, I hardly know how to put together a simple glue-the-pieces-together model available on the discount rack at the local pharmacy, with a person the ripe old age of nine could build. This motel kit is ten times more complicated than that. The wood is thin. Pieces need to be sanded, primed, and painted; the instructions are vague, short on describing the "easy" step-by-step processes, and have terrible visual aids.

Perhaps the biggest problem of all is - ME!

My patience is something that needs to be rewired.

My fingers are too big and calloused (two of them have almost no feeling at the tips due to injury.)

How the hell has my father done it? And how has he done it successfully? If I were able to call him - without disturbing him, that is - those are the types of questions I would ask him. Not, "Hey, how's the weather up there?" They would be more like: "How the hell do you keep your fingers from sticking to a piece of freaking trim board half the width of a toothpick?" Or, "Have you ever 'accidentally' thrown a bottle of Testor paint across the room?"

"This is supposed to be relaxing. This is supposed to be relaxing "

I whisper that to myself when things are starting to get frustrating. The irony: I started painting, putting together dioramas, and building models when writing became so frustrating; nothing good was going to happen by me continuing. However, especially in the middle of the night, I don't want to stop creating. My mind is still trying to function and be productive in some way. Sure, I could say the heck with everything: turn on the television and vegetate. But I don't. I jump from one thing to another. And, if all fails, when I know I'm done for, I'll clean a room in the cabin or reorganize my bookshelves.

Sometimes, I just need someone to talk to during the midnight hours. Someone to bounce ideas off of. Someone to learn from. Reading. There is always reading. And I learn a great deal from my books. Sort of difficult to have a dialogue with them, though. And I will admit it I have tried having a conversation with a book. I have argued and bickered with books and their authors. It's a terrible way to exist. Alone in a room, shouting at a book. Admitting to it - now THAT IS CRAZY!

Oh, Bless You, Mr. Vonnegut

It's 7:03 AM. On this mid-November Sunday, I'm sitting on my couch, thinking about Kurt Vonnegut. I'm reading his funny little book, *God Bless You, Dr. Kevorkian* (Seven Stories Press), a collection of fictionalized interviews Vonnegut had with dead people. What began as 90-second interludes for New York City's public radio station ultimately resulted in a book.

Most of the interviews begin with "today's controlled near-death experience" and end with "This is Kurt Vonnegut, signing off in the lethal injection facility at Huntsville, Texas."

(Kurt Vonnegut, Jr. died, realistically, on April 11, 2007, as a result of brain injuries incurred several weeks prior from a fall at his home in New York City. He was 84 years old.)

What I like about this slim volume of a masterpiece is the little tidbits of information a reader consumes about the past on interview subjects. For example, I did not know the famous American defense attorney Clarence Darrow (1857-1938) came from a small farming town in Ohio called Farmdale. I am from a little Ohio farm township, as well. Deerfield, OH, to be exact. Though I am far from being an attorney - a famous one, at that - I still feel a little connection and pride that I came from the same kind of grounds as Darrow and Ambrose Bierce (1842 - disappeared 1914), as well as Hart Crane (1899-1932). Crane was from Garrettsville, OH, in Portage County. Deerfield is a part of Portage County, as well.

Back to Vonnegut:

I started reading K. Vonnegut, Jr. in college. No. None of his books were assigned to me to read for a class. In a bookstore, one fall afternoon, on a display table of American classics, was *Slaughterhouse-Five*, Vonnegut's 1969 science fiction-laced novel about the bombing of Dresden during World War 2. By my fourth year of college at the University of Akron, I had read most of his novels and some of his short fiction collections. His style of writing both intrigued and inspired me. It still does. And I am not finished with devouring his work.

One of the great things about Great Writers is their capability to continue the element of surprise for the reader. God Bless You, Dr. Kevorkian, was published in 1999, and this was the first time I read it. More of Vonnegut's books will be included in my library. I'm sure the good author will bless me with some more surprises.

A Wet, Grey & Wonderful Day

When it is a grey and raining and slow day - like it is today - there is not much I can do on the farm. The only thing I seem to be doing is cleaning the paws of five dogs that, unfortunately, on this day, like any other day, need to go outside to do their business. And afterward, naturally, they want to come back inside the home to stay warm and dry.

They are also stinking, muddy wrecks of the Animal Kingdom when they come back in. There are paw prints all over the floor, mud on the new furniture, and a good deal of cleaning to do each time they come in. Leaving them outside is out of the question. That produces an even more annoying result: barking and scratching at the door. All of them coming in at once can be frustrating - a mega, hair-pulling pain in the ass frustration.

But they are OUR dogs.

They are my alarm clocks, my night watchmen, my therapists. Most importantly, they are my pack – my furry family members.

This will be the first thing I have written since moving into the new home. And what an adventure it has been. Fast and happy is all I have to say about it... maybe a little frustrating. Who am I kidding? It was very, very frustrating. But, at the end of the night before I go to bed with the wonderful and beautiful girl that is my wife, I MARVEL. I marvel at the fact this is my HOME! - That I am HOME! My books are placed, temporally, on temporary shelves in a room dedicated to my reading addiction. A library! My very own small library. (Tara and I will be making special, built-in bookshelves on the walls later.) My RAMBO knives and memorabilia are displayed in glass cases, complete with additional props and detailed settings.

We have a bed. I have a bed that is shared with someone so remarkable I can't even conceive of a God that would permit me to have such a privilege...

But he did... some things should not be questioned, I guess.

Goethe wrote: "Moreover, I hate everything that only instructs me without increasing or immediately stimulating my own activity."

Well, now, it seems like everything is increasing and stimulating me. It's a rare feeling for me to have the type of home and family again that makes me want to do well. To do something RIGHT and to be genuinely happy about the good fortune both Tara and I are blessed with.

Tonight: I will be cooking steaks and Gouda pasta for my girl.

There will be laughter. There will probably be some serious conversation about the future. What I do know is, in the morning, I will be in a warm bed with her, in OUR home.

And for those of you who don't know, it is the best thing in the world.

It is raining outside... I remember waking up in cold rain... in cold, grey, and terrible and horrible places.

There is no point in going back to that. And, dear Readers, NO ONE needs to go through that!

There will always be a warm bed for anyone who is honest, especially for themselves. Someone who will fight what is wrong may not always be rewarded, but the reward is – or should be - in gallant and selfless action. It can be lonely at times when you are in a battle. Never forget that. Also, though, it does not have to be experienced alone. More people are out there than you think. People that will stand next to you, wanting to help you.

I can promise you that.

Never stop fighting!

Book II

Most of these writings were constructed before I moved into the Locke Nest Cabin.

I was living in the small town of Taylorsville, NC.

Most of these pieces were to be published in the second volume of Notes from the Night Desk. At the time, when I decided I was through working on what would become the final manuscript to Volume 2, I promised myself not to be irresponsible and trash the pages I did not include.

So here is some work from earlier in the year, back when I was living on a farm and getting things prepared to move and sell the farm.

I hope you enjoy reading them as much as I enjoyed typing them up.

A Fun Part of My Day

It is feeding the goats that I enjoy the most on the farm. They are affectionate creatures. They sure as hell know when it is time to eat. I will fill up a large, red plastic Folgers coffee container with pellet-like feed. This feed is supposed to be loaded with nutrients, providing them with the health, strength, and good looks needed to maintain their worth.

From what Tara has told me, goat farming can be a lucrative business. I should hope so, but I never really understood. I mean, a person can eat a goat. However, goats are not something you see wrapped up in cellophane and stocked up in the meat coolers at the local grocery store. Nor is goat milk. Not like beef, cow's milk, chicken, pork… I guess I have never seen the advantages that maintaining and marking goats could have.

They seem more like pets to me.

As soon as I open the gate, they are after me. Jumping on me, tripping me. Three of them started rubbing the tops of their heads and faces on my leg.

We have the males separated from the females, of course. And, of course, they sometimes find a way to escape, make their way in with the females, and we have a helluva time wrangling them back to their designated area.

Like any other creature that is male, when it comes time to get a lady, we are just stupid, determined, and don't want to listen to one lick of anything anyone is telling us.

Reinforcing fences seems to be a reoccurring duty for me. I do not mind it, though. It makes me feel useful. On the other hand, I recognize the futility of this chore. – They will find a way. They will always find a way.

However, the goats are the animals I enjoy feeding the most. Not only do I feed them the nutrient pellets, but hay is a MUST, as well. And there is something about cutting the twine with my knife, the smell of the hay as I grab an armload of it to carry it to the feeder... the feeling it gives me... it makes me feel like a farmer.

Maybe that is what I REALLY am.

A FARMER?

My god, when did this happen?

A year ago, back when I was living in Ohio, I was managing a kitchen for a Sober Living home. Now, I'm feeding pigs, going to a feed store, fixing fences.

How the hell did all this happen?

Of course, I still read. I still write. I still cook.

It's hard for me to determine if "farming" is my job. I do not receive a paycheck; however, I receive all that I could ever want. All I must do is ask for it. It's that simple: if you need something, ask. If you desire something, ask. Don't steal. Don't lie. Ask. And I don't ask for much.

Those goats, though, make me smile. They attack me when I enter their pen with food. And today, while reinforcing the fence, no less, two of them were trying to help me. As I was slouched to nail-in a stud into a wall to secure a fence, one of the girls jumped on my back; the other was nibbling on the lanyard of my knife.

This may not be the life I imagined in my twenties... in my teens.

Hell, I could never imagine myself raising baby piglets.

(That is a story all of its own.)

What I do know: I have been blessed.

Why?

I have no idea. Maybe there is a God up in the heavens that feels some people have suffered enough... that their bill has been paid for their sins... that it is time for a break from all the broken moments of life.

Maybe I'm reading too much into it.

But I don't think so.

Transitioning & Cleaning & Packing

Having 12 dogs, of course, has its disadvantages. Spent time late this afternoon picking up torn cardboard boxes, feed bags, chewed-up gloves, and pieces of plastic to things my imagination can't begin to fathom. All of it strewn through the yard and fields. That does not even begin to cover all the "natural" trash/debris that covers the ground here: all the dead and dried and fallen limbs needing to be collected, put into piles, and burned.

We are moving. Well, we are having a new home delivered to the property. Tara has sold her smaller house. The newer one will be here in less than a month.

Getting things ready for the transition is something I NEED to be a part of. I NEED it for the psychological change I know the move is already causing. Both Tara and I have moved around quite a bit. Not together. This will be our first BIG, physical move TOGETHER. Yes. The move and transition from Cleveland to Taylorsville was very significant. - Hell, it was a SHOCK. For the first time in years, I felt safe somewhere. And I was able to be open. That my head could rest, finally, and BREATHE a little more freely, with the prospect of having prospects igniting more than just wishes that flash-burn comfort – that just exist to placate a person, never having a possible actuality – or reality – to them.

No.

Now, things and goals and objectives are being met and obtained.

For me, moving from one place to another is really no big deal. As long as Tara is with me, I really don't care. This little house, though, did something to me.

Setting up and working all the hours I do on the computer in MY SPOT has been VERY important to me. I had – well, STILL have - MY SPACE. And that means so much. I mean, I had everything I needed. Anything academic, to the most ridiculous things, like coupons for Adult Toys taped to the cupboard doors, were all there for me in my pursuit of becoming whatever it is I will become with the trot into middle age.

Packing all that up, which I did earlier today, felt odd... like another section of a chapter is nearing its finish. Which I know it is. But instead of racing off to the next location, not wanting to look back, I will want to remember my time here. It's a place I have no reason to repress.

I can say that I'm tired of moving. And all I want is to look at a building/house/structure and know - just finally KNOW - I will be there for a long while. And GOOD memories, laughter, and growing will happen THERE! That I will feel secure. No more Lost and Frozen Man for me. Time for some roots. And I'm trying not to be cold during the transition, and it's just normal for me to have possessions packed away.

I know I will see those items that were packed away today.

For so long, in so many places, I would not even accept anything that I could not carry around with me easily. Something that would not fit into a backpack forget about it.

Now, I have treasures again. That's right, as of this writing, I have STUFF. Stuff that reflects my personality, as well as items that comfort me. How is this so strange: one of my therapy items is an 18-inch Machete. Usually, that ugly bastard is on the wall to the left of me, within reaching distance.

(I wonder how many times during the day I reach over and walk around with that Machete, thinking, looking for inspiration... or, when I am uncomfortable, I'll have it in my hand, tapping the blade on my leg...)

I really wonder what the new place is going to be like.

I really wonder how much of me will change.

But I'm positive this is where I need to be.

Another One About a Dog

"Write me something," she said.

We were in bed, our arms around one another. - It was bliss, in the way people know what bliss truly is. And yet, my mind was not only focused on how wonderful my soon-to-be-wife was, but it was also on poor Coco.

Yes, this will be another dog column. What can I say? We have twelve dogs. They are a BIG part of our lives. Their well-being - both physical and mental - is one of the top things on Tara's list to keep in perpetual order and peace. Our dogs are our children. We raise and care for them as our own because they are our own.

These dogs mean more to me than my own brothers. They are, at least, kinder creatures and more forgiving and understanding. I think the fault of The Human is our ability to choose stupidity and prejudice over kindness and devotion. Freedom of Choice is not without its challenges and tests of character. Where a discontented human fails, a simple dog can prevail.

Poor Coco... she is not going to be with us for much longer. She is eight years old, a chocolate lab mix of some sort, and she is dying of the Big C. Cancer. And there is no such thing as a GOOD form of cancer; Coco's seems especially cruel. Either that or she is still so strong and determined in nature not to go out "quietly into that good night," as the drunken Welsh poet once penned. She is raging, raging, raging "against the dying of the light."

Last night was a rough night for Coco. She spent most of the day under the bed. That is not unusual. For as long as I have been here, Coco has spent the afternoon hours safe, tucked away in her private, cocoon world beneath the mattresses. During the evening, though, she tried to walk into the living room to get something to eat and watch movies with me and Tara. She was staggering, her legs trembling, and she was falling to the floor. Her struggle to stand back up was very real. And it was very heart-wrenching to watch. Tara had to pick her up to place her on the couch with us. Her head lay on her mama's lap, and her body was curled and shaking, nudged between the two of us.

Now, I'll admit Coco and I have not always gotten along. Sometimes, when Tara wants to get her off the bed, she will snarl and growl and show her teeth. That kind of behavior will NEVER sit well with me. She has an attitude problem. And for good reasons. I don't think I know of any living being in a "good" mood being in pain with Cancer. But, simply put, I will never accustomed to even the slightest possibility of Tara being in any danger… even if it's just empty threats from a sick dog.

(I must remind myself: she is in PAIN and dying. It is only natural the poor dear is no longer an exuberant, tail-wagging puppy.)

She was in pain. We gave her two different medicines for her. A couple of painkillers and a "special" cookie.

A good dog. She really is. And I hate the fact we will be putting her down soon. I have put animals down. Personally. If you know what I mean. But that is not what is going to happen between me and Coco, thank God. A veterinarian who does house calls will be coming at the start of next week.

I hope that is what comes to pass and that we can keep her at a comfortable baseline long enough for a professional to come to help her through the last thing she will ever have to do.

Last night was very bad: she wet herself while the three of us were relaxing together for the evening. We quickly got her off the couch. She could not walk. She stood there peeing on herself and on the hardwood floor. Coco knew what she did - what she could not control - and she was embarrassed. It was very sad. Tara and I assured her over and over again that it was okay.

"I'm sorry I can't fix you," Tara kept telling the dog.

(Indeed.)

I'm sorry about all of it, too.

It's always hard to lose a friend.

I'm just thankful she has lived long enough to have befriended me.

(AUTHOR'S NOTE: The vet never had the chance to make it to Coco. She succumbed to so much pain and bodily failure rapidly. It was too much to bear, thinking of what she must have been going through. I put her to sleep.)

Treasures

When you have twelve dogs, keeping a small house clean is close to being impossible. Tara and I have been showing our little house to potential buyers. Every day, the house needs to be cleaned and cleaned and then cleaned some more. All the hair and food and dirt needs to be swept from the floor, put in the trash, and the trash taken out before the dirt and dust and hair decide they don't like it in the trash can and pull an escape trip on the first gust of wind, or ride from a breath of breeze.

Every day, dishes need to be washed and put into the dishwasher... dirty laundry needs to be collected, shoved into the machines, washed, and then put away.

And let's not get into my work/writing space. That always remains a wreck of Post-It notes, stacks of manuscript pages, cuttings, open books with highlighted, helpful passages, and various print-outs covering various subjects… all of which span a very limited amount of space.

Windows and doors need to be opened. Fresh air and air fresheners need to cut away the smell of dog and cigarette smoke and the general scent from our everyday lives as humans, secreting unique scents from oils and sweat. The smell of the US, more or less. Our body oils mingled with soap and cooked foods. The smell of a home.

Right now, the house is ready for a showing. Our new home is bought, paid for, in full, and waiting.

Tara is doing some laundry. I'm simultaneously writing this piece and reading an excerpt from Barry Miles' biography of Charles Bukowski, a book I received in the mail today. I'm running on this trip to replace the books I lost to flames in Ohio. First, Editions were packed away and stored. They succumbed to Black Mold; hence, I put into a bonfire that licked up and burned the pages as fast as the tears ran down my face.

At one time, I had over a thousand titles. Dozens of writers. Writers such as Norman Mailer, Joyce Carol Oates, Hunter S. Thompson, Jack Kerouac, and the like. Some editions were signed.

Many things were lost when I abandoned life. When the divorce happened, all I cared about was getting drunk. Black-out drunk. Not caring about the treasures I had spent years hunting down and collecting. Not really care about anything; that was me for too long.

Now, I am digging and searching for those treasures again.

Whenever I get one piece of the lost library back, I promise myself never to abandon what I have ever again. My life is with Tara. We are building a world together, filled with all the things we could ever want.

One day, I will not want to ever leave home. Why would I? All I will ever need – friendship, safety, love, a sense of being where I am supposed to be – will be right here.

Polar Wolves

Growing up in Northeastern Ohio, I know snow. Everything like the Lake Effect, ice slides, moist and dense and heavy snows, bursting water lines, and deadly six-foot long icicles spearing their way down from the edge of rooftops. Salt sprinkled on the roads and chains wrapped around tires were part of an everyday waking morning. Coffee-drinking conversations sprinkled the talk of road safety at the kitchen tables across the region.

Schools shut down for a day, maybe two days at a time.

Generators need to be purchased, along with five-gallon cans of gas to fuel them.

Firewood stocked.

Snowball fights abound in the Ohio countryside. Melting and hard-as-rock frozen snowmen in various morphing stages ranging from the magical to the grotesque.

That was my Ohio upbringing.

Later in life, I moved to Montana. And, dear Readers, when it decided to snowstorm throughout that wonderful state, it not only white-out STORMED, it blew the cold-hell blowing of winds that convinced you a God was real. And, on that day, that week, that month - He was pissed off at something!

Now, I'm in North Carolina. As of this writing, what is going on right now? There is a winter snowstorm.

RIGHTY!

A snowstorm!

From someone who came from Ohio and lived in Montana... this is no storm.

If anything, it's just another mild January winter day in Ohio.

No buildings are caving-in upon themselves from the weight of snow. No one is freezing to death on the streets, in the alleys devoted to the homeless and deprived. No frost bites. No shivering and starvation, with flurries freezing to eyelashes, the body so cold it forgets what any other feeling is but COLD and PAIN.

This is no Storm of the Century, the terror of wet and white flakes of accumulation. It is just SNOWING.

My wife, Tara, has four large, 18-month-old puppies – Great Pyrenees. They are the size of teenage polar bears. No joke. As far as the canine categories go, they are a large breed. They are supposed to be protectors, especially during the night.

But is that what has happened since this snow has been falling?

Oh, no.

They have been puppies. Rolling in the white and pure, dirty powder. Barking at nothing. Running wild. Jumping on one another. Wrestling with one another. They have been fun to watch.

Ms. Tara – Mrs. Locke - is calling them her "Polar Wolves," which I, stupidly, tried to correct.

"Don't you mean Arctic Wolves? Or North American Wolf," I chimed in, trying to pretend to be intelligent.

She was doing her hair at the time. Looking in the mirror, spraying sweet-smelling spray into her wonderful long strands of hair. Brushing and straightening it.

"No. Polar Wolves," she corrected.

There was no point in arguing.

We have Polar Wolves...

And the South has no idea what a Blizzard REALLY is.

For the Sake of Argument

Let's say I might become successful at something. Something like Spelling and/or Grammar. I fashion myself as a writer. But I am horrible with the RULES and STRUCTURES of writing. What would happen if I followed all the rules - knew all the rules - and applied them?

Would the work be better?

It would be cosmetically, politically, and structurally correct. But I cannot help expressing that something special would be lost in the piece. Whatever that piece might or might not be. Whether it be a column, a novel, an essay, or a poem, I understand rules need to be applied more to things, say, an essay for a college course or a job résumé.

I have never been good with Rules. However, I have not broken them, most of the time, to be a malicious person. Or a lazy person. I have been possessed more by Dionysus, God of wine and insanity, rather than Apollo, God of the sun and of sober order.

Even with the changes I have made in my life in the past year, I have not followed all the rules. Humans - all humans - are fallible by nature. And some are more defiant than others.

I'm one of those MORE defiant types. A hungry and restless and pushing type who, from some freak switch that has been flicked ON, has a different kind of brain chemistry than others. The doctors and counselors agree with me on this assessment. I even take medication for it.

And that is no smoke being blown up your bottom. No, no. More than one doctor, more than one therapist, has said that to me: that I am "wired differently," which can be a good thing. On the other hand, it can be the worst thing about me.

Let's go back to the original question: What would happen if I became successful in some way? And I am talking about the kind of success that is considered Socially Significant. The fact I write every day, I consider that an accomplishment - a small, personal success. But I have not held down a paying job in years, Child Support payments… I don't even want to get into that. I have legal fines I need to take care of before I can get my Driver's License back.

On paper, I'm a mess. A failed, irresponsible American adult male.

Internally, though, how I have been feeling about myself these past few weeks is far from being a failure. A success? Not yet. But I'm getting there.

With each day that goes by, I don't hurt or bother anyone, which is a wonder; each day, I dedicate myself to working on The Word, and each day I remain with my wife, I feel successful. - Even a little bit rich!

Maybe one day, thinking of myself as a loser/let-down will abate. I'm sure as hell better off than I was yesterday and the day before that, and don't get me started on the years ago, especially three years or four years ago – five, even!

The Appt., Revised

(AUTHOR'S NOTE: This piece was started in the first two weeks of my arrival in North Carolina. Since then, I have had several visits to the Doctor and with therapists. All without ill endings. It does not get easier to go to see these professionals. But I want to be - and remain - better… if not a little damaged.)

What is it about going to a doctor's office?

I get nervous, not about them finding things wrong with me. Hell, I KNOW things are wrong with me. The depression, the horrible ADHD with its "unfocused" curse and "stuck" side effects, and the anxiety (exacerbated by anyone wearing a white coat and pens in their pockets).

All of it!

So, yes, I have a doctor's appointment today.

Why in the hell am I doing this?

I am doing it because that Special Woman in my life wants me to do it.

She wants me to be healthy. She wants me to have the right Medications. She wants me to be my best.

How long has it been since someone cared so much for my health? Someone who was actually, truly associating with me. BEING with me!

Let me put it this way: There was The Red-Headed Demon. She did not like my drinking. She knew I drank too much - WAY too much!

Her bright idea: Stop Drinking! Seems logical, doesn't it not? But I liked the BUZZ. Any kind of buzz or high or whatever you want to call it to kill the feeling of feelings.

"Just smoke CRACK, like me," she said.

That's right! That was her answer to my drinking problems: smoke crack.

The thought of it made me feel wrong. Like I was on some higher Under-the-Influence moral ground, I would drink mouthwash, for Christ's sake, to keep from getting sick, making me sick along the way. The minty-flavored buzz I got from it was not that bad. But, my God, that will make you sick in its own individualistic way.

(This is something I will have to tell My doctor: my drug intake history. Not looking forward to that. And it's a subject I NEVER look forward to discussing. Same with drinking. I'm over it! Don't have the patience or interest to rehash how horrible some of those times were. And I was forced to talk about it extensively. It feels like that bad movie that always seems to be playing on the same channel. And, for reasons unknown, we watch it.)

Cheap and Dirty and Wrong are finished.

For me, at least. I hope, anyway.

Time to feel good. My medications... my supplements... my relations to those I love... talking to you, dear Reader, is a far better thing than anything I can smoke, drink, or snort.

Growth is supposed to be hard and painful.

Pray for me that the Doctor's visit is not as bad as I am making it out to be.

Uncomfortable?

Yes.

But I can do it. My support will be right there, next to me, holding my hand the whole time through. I know her well enough to know that.

And, on the bright side, I've already met the Doctor. Socially, that is. He is a friend and a former coworker of my girl's.

Everything has been falling into place well here in Taylorsville, North Carolina.

What in the hell took me so long to get here?

Shifting Gears

One of my biggest problems is being too out of touch with everything, of becoming too comfortable with isolation away from my generation - a generation I, at one time, felt strongly compelled to write about and define. So many Beat Generation writers dedicated more words to detailing the intricacies of friends and lovers, detailing what would be the classic texts of an original movement in American literature and culture.

I'm envious of that. Envious in the sense there were enough of them to be a movement. Envious of the material they had to work with.

Isolation is something I am good at. It's very uncommon for me to associate with anyone for any length of time. My girl, Tara, is always with me. And she is my true companion. My true best friend, me being isolated, does not seem to bother her as long as she is with me on the trip. Yearning for more social activities is something I will not deny, and I have been wanting to do. People do interest me… well, at times, and it depends greatly on the other person whether I am comfortable enough – not to mention interested enough – to constitute my time with them.

Charles Bukowski wrote in his second novel, Factotum: "Isolation is the gift. All the others are a test of your endurance, of how much you really want to do it. And you'll do it despite rejection and the worst odds."

You could not be more correct in your assessment, Mr. Bukowski. I've kept on doing it. And most of the time, I enjoy my isolation. Pushing myself to be more active, though... that seems right, as well. Never have I been one to make friends easy. I'm talking about the REAL kinds of friends that are genuinely GOOD to one another. That is hard to find in the Addict's World. Sadly, that is where I have spent a good deal of my time over the past fifteen or so years. Five of those years were spent amongst criminals. It does not make for a good field for friendship tryouts.

It does not make for a GOOD field for anything… except guaranteed pain, loss, and deconstruction of personal and social morals. You know, the good stuff.

It's not like I went out to dinner with those people. That would have required money. If there was money - ANY amount of money - it was spent on the demons demanding to be fed within us. All the men and women in the rehab institutions and hospitals I was in were Associates. There were no REAL friends there. We were forced to talk with one another. Forced to eat next to each other. If I liked someone more than another, I spent my time with them. But it was not going to blossom into something that would endure.

I did not want that.

What I would like now is to be around people who are the polar opposite of me. I mean, I would not want a friend who could be a possible risk to the serenity I have acquired. If there is going to be a problem, I want to be the problem.

Hopefully, I'll find some people out there who will put up with me. The hard part, at my age, is time. We all work, spend time off with family, or personal pursuits… hell, I can count on one hand the amount of visiting "friends" that have been in my and Tara's home.

It is conflicting: wanting and being isolated and desirous of some company now and then.

Getting Out

One of the interesting things I'm currently going through is DOING things. Things I would have never done before. Tara, who does so much, does not insist that I do anything that forces me to struggle with an unessential task. She just supports the choice I have made to expand my world a little. And stay as healthy as someone like me can be.

For example: We have a place. And I am not talking about our home. I'm talking about an establishment. A restaurant called The Vault. The building was formally used for a bank, and guess what: the bank vault is still in the building. Hence, the name. We would like to be dubbed "regulars" at The Vault. We want to know the servers, the owner, and even the people who perform live music there. And we would like them to know us.

Why is this important to me?

I really don't think I have a proper answer. I just know it feels right. That feeling of being a part of something again is an amazing thing. Even if that "thing" is just being a regular at the local eatery.

But enough of that gibberish, for now.

Today was a GOOD day. Not that I have had any BAD days during the weeks I have spent in North Carolina; it just seemed like today was an exceptionally positive and constructive day. (I have not had a truly BAD day in some time.) For one, I received more "Likes" on my blog today than on any other day since I started it. True: I may have started the whole thing up back in 2013. But it did not become an essential, daily task to Post any form of work until the summer of last year.

And it feels good to know I may be reaching somebody out there with these funny little sketches. That I may, dare I say it, actually have readers. Grasping such a concept still feels foreign to me. Almost overbearing. See, a great fear of mine is humiliation. And, for the love of God, if I were to make an ass out of myself in my writing... well, I wouldn't be able to forgive that.

Another important thing that made today feel so good was how much I got done in the house. I have not mentioned that my girl and I are moving. Well, it's not really moving. We will still be on The Compound, but the house we live in is up for Sale.

Our first showing will be this evening. It took less than forty-eight hours before people were messaging and calling about this little place. AND WE ARE NOT PREPARED!

There is all the packing and cleaning and organizing and throwing out what is, technically, not of value. She has lived here for ten + years. Enough time to accumulate MORE than what is needed to survive. Me: I arrived here eight or nine weeks ago. When I arrived, I had two bags and a sad lamp I refused to leave behind in Cleveland because it was the first piece of furniture I had bought in years. Sure, I have more material possessions now than I have had in five years. But not so much a good hour and a couple of boxes wouldn't cure to erase any evidence I so much as even walked through the door, let alone lived here.

Maybe part of that is the problem. I have not left a mark on any one place I have lived at for so long to leave a mark. Packing what possessions I have now seems sad… almost as if they were never here or meant to stay.

It means it is time for me to leave again… without leaving that mark.

Don't take this reading the wrong way. Both of us will be moving to a bigger and better place. Together. There is even an alternative place to stay if we do sell the house soon.

However, there is a sting to the situation of me packing my things back up. Getting settled in. I can't imagine what it feels like for her. - This place, where I write these very words, this was/is her home.

Anyway, I moved a lot of things around today. Cleaned. Took care of the animals. She and I had a good afternoon and evening together.

I thought to myself, as we were passing one another from one part of the house to another: Maybe she is my home as long as her door is always open to me, which is where my home will be.

What a Day

It was 1:48, exactly, on this Wednesday afternoon in North Carolina when, once again, something hit me. I'd been awake since five - I'd gone to bed at a little after three in the morning. I'm still burning the only brand of oil I know how to burn: the kind granted to me by a doctor. Not much sleep, but some. Anyway, that was not what had my nerves jacked up at that very moment in time.

1:48 pm. My mind could not recollect those numbers or that time was associated with anything demonic or disturbing. But for me, at that moment, they were supernatural digits.

Maybe I should back this story up a little. Tara, I now can call her by her name - TARA - instead of My Special Lady Friend (and that is part of the weirdness the afternoon has brought, as well,) and I had to be up early. We had a house to go see. A potential home. And I knew I was not going to be myself in the whole process of discussing improvements to the heating and electrical complexities of a home built 22 years ago.

I just don't know a damn thing about it to say anything. So, I don't.

And there was the truly terrifying question posed to me on the ride there: "What do you want?"

What do I want? My god, I never had an opinion before today on what I wanted as far as living arrangements went. What was affordable? What place will it take me? Or, in really, really desperate times, where was there a bed open or a space on a floor I could occupy?

There were a lot of things I should have been thinking about on the ride to the house. I imagine common questions, such as a fast push into the next step of the relationship Tara and I were embarking on, should have been posed. But that was not on my mind. There is an unquestionable confidence I believe exists in me regarding us - Me and Tara - which should just be left alone. - It should just be accepted.

Maybe I should have been thinking about water damage and drywall replacement...? Perhaps the possibility of the roof needing repairs…

But no.

It was the email I opened and read two hours before Tara was even awake, pulling me away from the NOW of the moment. It was an email from my Project Manager. A nice woman named Jen forwarded me the first images of my book cover - front and back flaps - along with the forms I would need to fill out if I thought changes needed to be applied to the layout, the image, and the text.

It did not matter I had under two hours of sleep. I spent the next two hours, before Tara even sipped her first sip of coffee, meticulously - almost obsessively - finding flaws that needed to be corrected. Sentences that needed to be reworked. And that is not a part of who I am! But this is a book, damnit, MY first book, to boot, so I congratulated myself by being an amateur prima donna, making the product worse and trying to improve it.

I needed to breathe. Smoke a cigarette. Get out of the house. Go look at another one. But not before I filled out that damn form, attaching to it a Word doc. Full of my concerns and suggestions on how I want "The Book" to appear visually.

And what do I know about Graphic Design?

About as much as I know about granite countertops.

In a way, it was starting to look like a screwed day. One filled with my personal inadequacies and irritable pathology.

I mean, the first cover art for my first book was right in front of my face, and all I could think about was how to make it better! And how it was not how I envisioned it. Lord knows how I will be rereading the whole thing before it is printed, looking for flaws - or, I should say, what I consider flaws - wanting to expand certain sections... shorten, or completely cut others. I'm going to be a wreck.

Sleep will become a diminishing pastime.

Back to the home, though. It was nice. Solid. Had a good feel to it, though the situation the current owners are going through is bad.

We like it. Tara and I were interested in everything that was being told to us about the house - everything that had been WRONG! And about the things that had been fixed during their 22 years there.

There were some cosmic, yet comedic, lessons going on. Yes, I was picking up on its not-so-subtle method of delivery.

Before we left, we put down a cash deposit.

Now, for those of you who have not been reading my posts during the past year, I am broke. Probably will always be broke. I have enough to keep myself from starving – that's about it. And enough to smoke cigarettes. I'm proud to donate to the Gonzo Foundation every month. I have enough clothes and books and everything else. And I'm grateful for it all.

Terms like financing and mortgage and percentage rates are more terrifying for me than the worse DTs hell-ride into terror. Tara, on the other hand, is a genius when it comes to such things. And, because of such things, she has become quite the sardonic whiz when it comes to the Almighty Dollar.

When I was told the price for the home, I was not sure whether to be happy, mortified, disappointed or frustrated.

(I never bought a home! I don't know what the hell to think. Put me in a room where people are discussing the significance of Hemingway stating John O'Hara was the most important novelist of that generation; I'll have plenty to say - plenty of thoughts would be percolating in my skull... but this domestic, Strange Stuff is out of my weight division. A man needs to learn the hard way when NOT to step into the ring because he remembers what it was like to tap out... remembers what it was like to be knocked the hell Out.)

So, the next stop is the bank. But not for what you are probably thinking. Tara had to get the separation papers from her husband signed and notarized.

(That is now why I can, without any legal ramifications and strain, call Tara by her name instead of the mysterious My Special Lady Friend.)

It wasn't until we had pulled out of the Bank's parking lot that I acknowledged the question that had to be asked - that I needed to ask – had to come out: How are we going to pay for the house? The house we are living in has not even Welcomed so much as her first potential buyer, and we HAD OUR NEW HOME WITHIN 72 HOURS!

Madness.

What was crazier was the phone call Tara had with the bank. - She was just in the bank seven minutes prior to placing this absurd call. The call went something like this:

Tara: Hello. This is Tera T__. Is Darcy there? Yeah, hey. Yeah, I was just in there. Nope. Didn't forget anything. I was just wondering how I could get $XX.000.00. I just got a home.

(Indecipherable noise from the iPhone.)

Tara: Yep. It'll be for the farm.

(More indecipherable noise.)

Tara: So, I'll get a checkbook. And how much can I ask for? I mean, I don't know. Anywhere between $10,000.00 and $2,000,000.00! Wow. Well, I think $ XX 000.00 should do it. And I pay that back by the end of the year? Um, when will the checkbook be ready?

(A few seconds of agonizing silence.)

Tara: Ok. That sounds good. You have a good day, too. Yeah. See you then.

She pushed the button, putting the iPhone into a wonderful snooze.

"Well," I asked.

"We can get the checkbook next week. I got us a little extra money, too. For moving expenses."

My mind had mushroomed grey, unbelieving brain matter. That phone conversation lasted less time than it would have taken to have a healthy and enjoyable bowl movement, reading something interesting during the process. No showing of photo identification. No forms to fill out. No NOTHING! And she gets enough money to pay for a home - PLUS EXTRA!

My consternation and utter amazement must have been as visible as if someone kicked the dangling joys between my legs up into my stomach.

"It's small-town living down here," she said with the Million Dollar Smile on her face as if that was supposed to EXPLAIN such insanity and recklessness. And from a bank, no less. Hell, when I go into a bank, two forms of photo identification are required: my mother's place of birth needs to be memorized and spelled backward, medical records are researched, and if the teller just so happens to be female, my medical history seems to make them blush; I have to swear an oath to donate more blood... - point being: no way in a sheetrock royal hell will ANY bank give me ANY money, let alone the amount Little Ms. Tara got with a little length-of-a-fart phone call. A bank won't even withdraw MY OWN money from MY OWN account without identification. Even then, they are skeptical.

By 1:48 on a Wednesday afternoon in North Carolina, I helped pick out our future home. We have the financial backing to procure the above-mentioned home. No longer does My Special Lady Friend need to concern herself with her soon-to-be ex-husband.

And I got to see even more proof that my book was going to be published.

(Something I thought would never happen.)

It was pretty cool, too. This whole day of changes… GOOD changes. Even though I have an odd way of expressing it, I am in awe of this life and the woman I share it with.

Record Broken

Something amazing has happened today. Well, it's amazing to me. It's 4:07 in the afternoon. I don't know how many times a day I check the status of how many views I get on WordPress. - It has become an obsession to look and to know I have people reading my writing. And knowing how many people are reading/viewing my NOTES FROM THE NIGHT DESK blog is a part of that obsession.

Today's NOTES... has had more readers and viewers than any other day in the blog's entire short history: 162 Views!

It's the first time I had over 100 people reading my work - all on the same day. And there are still hours left before the cut-off happens, which I'm not sure when that is. What I do know: it's before midnight.

When I first started blogging, I did not think much about it. Did not think I would actually get readers. I mean, maybe some family members... a friend, maybe two. Now, this was before I got a fast & frustrating education on Social Media. And all the different facets and links Social Media has connected to one another. For example, I did not know I could SEND and SHARE my blog writings to different websites. For years, I was very anti-Facebook, anti-Twitter, and anti-everything digitally published.

It just was not attractive to me. I was a writer. I wanted to write. I wanted a publisher to accept the work, manufacture a book, and then the book would go to stores.

It was a narrow mindset. I'll admit that. I was stubborn. I did not want to learn what was in vogue, what EVERYONE else was doing. Being against things comes as Second Nature for me. Why that is, I don't know. But how things change. When I saw the high number of views (high for me, anyway), I was shocked. I was a little weirded out, and if we are being honest, GRATEFUL!

Grateful people were taking the time out of their day to read some silly blog.

Why aren't they reading books? My Old Brain would have screamed. Why blogs? That's NOT writing. Those are people who aren't REAL writers!

I'm serious.

That was the 25-year-old me, mad he was not being discovered by some publisher or agent in New York. In two days, I'll be 38. I have more people reading my work now than when I was 25.

And guess what?

I'm not as angry.

Oh, But They Keep Us Safe

Okay! I will admit it. Sometimes, I do not like them. As big and as fluffy-looking and puppy-like as they are, the problem is that is what they are: too big, too fluffy, and immature pups. It's not that they are in the way so much (even though they are), as we all live in a house too small for all of us. Especially THEM! The Polar Wolves, as Tera, has so poetically named them.

The worst one of them all, though, is the one we call Big-Big. And that's truly what he is: BIG. One really, really BIG dog-puppy, with the face of an already persecuted martyr and eyes of sadness and longing. He is the largest of the four wolves. The funny thing is that with that face of his, he is the least intimidating. His face shouts: "Love me. Why aren't you loving me?"

Willie is his real name, named after Willie Nelson, I believe, and it's not his size that is so bothersome. It's that damned, sad temperament and looks. All an intruder needs to do is look into those big, sad eyes and realize there is not one vicious bone in his big, furry-white body.

The Pyrenean Mountain Dog is a breed of livestock guardian dog from France, where it is known as the Chien de Montagne des Pyrénées or, more commonly, the Patou. It is called the Great Pyrenees in the United States. The breed comes from the French side of the Pyrenees Mountains that separate France and Spain.

And they are big!

We have four of them. Two males, two females. Each one is named after a famous Country Western singer. Like the one female, June, as in June Carter. Hank: Hank Williams. And then there is Messy Jessy. For the life of me, I don't remember which song Jessie, our big Messy, is revering, but I hope the singer was/is not as defiant and downright dirty as our North Carolina girl is.

Messy Jessy has to be the middle child. She has the temperament of one. And I know what I'm talking about. I'm a middle child. Something about the middle child defying authority, causing problems, and generally doing it in such a way everyone knows it is not out of malice - it's just how we are.

Yes, I'm aware I am comparing my own proclivities to that of an eight-month-old puppy the size of Alaska.

Maybe that is the reason I like her the least out of the four.

Now, I don't want this to read like a hate column against dogs, specifically the Pyrenees, because I do like them. Most of the time.

What has spawned my interest in devoting an entire piece of writing to them has to do with what is going on right now, in the middle of the night, as I am typing this very column. The Great Pyrenees is a nocturnal canine. Natural protectors. These dogs had the training to be livestock guardians, I think. Plus, it's just in their nature.

How do they begin to protect if they think there might be trouble? They BARK and BARK and BARK. What makes Big-Big the worst Pyrenees out of our four? He is so brave and big and ferocious he feels he does not need to leave the front porch, sometimes, to bark at whatever imaginary danger is out there in the dark. 200-and-some acres of farmland and forest to be free to roam upon all night… Big-Big chooses to stay close to the front porch, where there is a security light! A blanket. And, possibly, if there was real danger, I imagine he wants to be the first one let in so he can hide under the bed. That is no stretch of my sometimes-excessive imagination. I have personally witnessed that dog - that great big, ferocious guardian dog afraid of something and moving as fast as his clumsy, puppy-dog-male-child body could go beneath the master bed.

That is our Great Protector, I thought. Our Great Guardian!

His baby sister is more of a threat than he is. All he has going for him is his size. And one hell of a loud, non-stop bark.

Maybe that is why he won't stop? Maybe he's worried if he does, the Great Beasties of the night will come and get him. If that is the case, I wouldn't stop barking either, I guess.

Plus, all four of them are just Polar Wolf puppies. They'll grow into those tall, fluffy bodies soon and mature. Jessie might learn that when her collar "reminds" her to get her hindquarters back on the property, she should really listen to it. Otherwise, she will get a very special electrical kiss on the neck. Tara's cowardly wolf might find his courage. The other two... well, they have not left the kind of impressions their big brother and messy sister have.

They are good pups. Hopefully, I won't ever have to protect them.

My Visit

Last night, the Menace visited me. Tara was asleep. The television's volume was low. I was not trying to sleep, not trying to stay awake, either. I just lay there on the bed, thinking, this is my birthday. My 38th birthday! I made it another year. I was thinking about the past year, WHERE I was a year ago, as to where I am now when it spoke.

Why books?

That is what it likes to do: taunt me. It taunts me by questioning my likes. My dislikes. - Everything.

Turning from one side to another, trying not to wake Tara up, trying to move around her sleeping body, I still hear its voice. Still, try to shut it out.

Nobody cares what YOU know! They care about what they can get. And that is all. And they can't get ANYTHING from you. You have nothing that is wanted. Throw in the cards, old sport. Give up before you get TOO old to retreat. You are getting TOO old.

38 IS a big deal to me. 39 will be a big deal to me. 40 will be a sweet, maltose miracle - the big FOUR-0!

Who would have ever imagined I would make it THIS long? 40 would be miraculous. The beginning of the best years of my life.

I'll still be here when you are 40. You will never get rid of me.

It is standing in the corner of the hotel room. I do not bother to look at it. I know where it is, I know what it looks like, and very well know what it smells like, like meat rotting in a puddle of putrid water. Rank and damp-smelling, it is. No reason to open my eyes, no reason to argue with it. There is no point. It wants my attention, that is all. That is one of the reasons it smells so damned bad: to get my attention.

That is all it ever wanted: attention from me. And influencing me.

I'll be with you till you are dead.

I hear it breathing. I'm not afraid. Haven't been afraid of it for years. And it is right. I know it is. It will NEVER go away. When I do listen, when I do BELIEVE its truths and lies, you might as well put me in a can and bury me with the dogs and cats and aunts and grandparents.

When I listen - when I believe it - all manner of DRIVE is drained and tapped out of me. My arms won't lift, and my legs move like two dangling, over-cooked spaghetti noodles, waving their way from the fork of life.

Stop trying to be poetic!

It hates attempts - AT ANYTHING! More and more, especially as I get older, it tries to be louder. As I get older, I hear less.

It is standing there, in the corner of the hotel room in Asheville, NC. It will be waiting for me back in Taylorsville. I know it's there. I know it will always be there.

And I'm fine with that.

It HATES that I am so fine with that.

Roadside Ditch Pup

Finding a treasure in a ditch alongside a country road may seem out of the ordinary. Last month, I sure as hell did not expect me and my wife, Tara, to find one

We were in South Carolina, visiting her brother and waiting for Grandma Vicky to join us during our brief stay. I had never met Grandma Vicky. She had read some of my work, though, and nothing gave me any indication of disapproval of me marrying her granddaughter. Family is family, nonetheless, meaning reading someone's ridiculous columns is no "first impression once-over."

All of us - Me, Tara, her brother, and his lady - were waiting on the back porch for Grandma when a phone rang. I heard my brother-in-law speaking, assuming he was talking to our expected guest. And he was. When the phone was passed over to Tara, I could tell something was up... nothing bad, but there was that feeling in the air, that feeling when you know something is about to change. That our little gathering was about to get a little more interesting.

"Grandma found a puppy and its mother. We are going to help her rescue them."

Oh, boy. This isn't happening.

My wife, loving her as much as I do, suffers from a rare and chronic condition of Caring Heart Syndrome that, consequently, leads to addiction. This addiction is best defined in simple terms: she is powerless against helping animals in need. Specifically dogs. And puppies are the "good stuff," as they say on the streets.

As soon as I saw Tara picking up a light brown fluff ball with a black face, it was over. Instantly hooked is the most accurate way to put it.

Now, she said it was my call on whether or not we keep the pup.

ME!

Right.

I knew I had to play that hand right, not just for the health and safety of the wiggling fluff ball in my wife's arms but for MY health and future happiness.

Naturally, I caved. Not only was the look on Tara's face something to make your heart stop, but that little pup was just so freaking cute. The blue-tinted puppy's eyes... the smell of puppy breath... a natural disposition of not knowing how to move the body in any way resembling grace... feet too big - ALL of it so damn cute.

We named her Winnifred, a.k.a. Winnie.

It's been about four weeks since we brought her home to the Compound. She still does not know how to walk in any graceful way. All that wiggling from her rear end throws off any possibility of coordination. Her mouth is filled with needle razor teeth. These teeth seem to need to be used every second of her waking moment. And for some reason, toes and fingers seem to be the top-shelf choice for little puppy dogs.

Do I like this fuzzball that seems to be growing at an alarming rate? (She was 5 lbs. when we got her; she's now 13 lbs.)

Sometimes I do.

Sometimes she is just an annoying little/big turd. One that is VERY demanding of everyone's attention. Not just Tara's and my attention. Attention from the other dogs, as well. That has been fun to witness. Most of the other older dogs seem to lack the patience for wiggling and bouncing and barking little things brought home from a ditch.

She is a treasure. I'll admit to that. We believe she is an Anatolian Shepard, a breed originating from Turkey. A BIG dog she will become. Like our four Polar Wolves, the Great Pyrenees, Anatolian Shepherds guard flocks of livestock and property. From what I read on akc.org, the Anatolian Shepherds are mentioned in the Book of Job. A story dating back to 1800 B. C.

It wasn't until the 1950s before this breed started to be imported into the states by ranchers wanting the dog's fine qualities of protection and independence. However, in the late 1930s, one male and one female Anatolian were given to The Secretary of Agriculture, Henry Wallace, by a Turkish ambassador.

It was a failed experiment to see what breed of dog would make the best sheepdog.

Whatever she is, she belongs with us now. A pain in the butt she may be - I wouldn't have it any other way, though. More smiles are spread in this home during the days because of her.

She will, as I know, become a regular personality in these stories. So much time is spent trying to keep her out of trouble and in control, and I HAVE to write about it.

Ornaments Are Meant to Be Put Away

It used to mean so much to me. I wasn't sure if I was covering myself with all those rings, necklaces, and bracelets. Was I covering myself up or decorating my body into a leather and silver Christmas tree-like art piece? I'm not sure anymore.

I would have six or so necklaces dangling at once, wrists loaded with bracelets - and RINGS! Oh, the rings! Those were my favorite. Small, ornate rings; large and hinged rings. Rings that seemed more like medieval claws and spikes - on ALL my fingers. Even my thumbs.

It was a glorious expression of excess and absurdity. But I took it seriously. Then came the day when I didn't grow so much out of it; I just didn't see the point anymore.

An external presentation can only go on for so long and so far. Identity has more to do with actions, how you treat others, and how you treat yourself.

Nowadays, I'm VERY "boring" with my look. Usually, grey or tan jeans, a tee-shirt with a button-up work shirt over it (all articles of clothing a solid color), Ariat Western boots, and the only "ornaments" I wear are my plain, gold wedding band, a SOG pocketknife, and a Luminox Navy SEAL/Leatherback Sea Turtle watch.

Everything a guy like me needs for an average day in the country.

Somedays, I remember to hide a pen and a little notebook in one of my pockets. You know, just in case...

It's the wedding band, though, that gets to me. In a good way, that is. I find myself looking down at it more and more as the days turn into weeks, growing into months, my thumb spinning it around that special finger, remembering its significance. The finger contains that one vein that leads right to the heart. A tradition and belief going back to the 11th century developed from the Roman anulus pronubis. A man would give a ring to the woman at their betrothal ceremony. Blessing the wedding ring and putting it on the bride's finger.

Now - after going through dozens of rings throughout my life - I finally have one that is meant to be on my finger. And there is one powerful vein running through me - all the way to my heart that refuses to stop beating.

It's tried a couple of times.

But,

It changed

Its mind.

Some Things I Just Won't Take

When someone I love or admire is attacked, my first obligation is to the victim. To assure them there are many cowards in the world - the type of people that hide behind many masks & platforms and spew vile, immature hatred into the world... to hurt, period. And yes, I do believe people who are verbally or digitally attacked by pathetic maggots with malicious intent are victims. All people have thoughts, feelings, and reactions to others. An online world is just as much of a world - maybe even more so - than Our physical world. I can't randomly start preaching and insulting and screaming insults to a stranger on the street without repercussions. My goodness, you would be in REALLY big trouble. Just imagine if someone took a real offense? What if someone hit or shot at you?

What put me in this mood this afternoon? What is driving me to write this piece?

When I woke up, I checked my messages and Facebook newsfeeds. On my feed was a horrible rant of such disgusting depths of stupidity towards a wonderful Southern man and writer - a man I admire. I was physically ill after reading it.

Accusations of this man being the type of man that is the main problem with society... that he is a misogynist... "probably a bigot," and "[he} represent[s] white male privilege and are thereby a microcosmic representative of everything that is wrong with this nation."

Well, I would like to retort to this ignorant, shit-filled, arrogant motherfucker.

I was introduced to this wonderful man's writing - this man you are trying to break down with hate - at the lowest time of my life. A time when I SAW what some of the TRUE MEAN and DANGEROUS aspects of this Nation were a waking part of my life - a life surrounded by people that PHYSICALLY HURT others and stole from others - FOR FUN! REAL walking, diseased, addicted devils... people that had no empathy or sympathy; no patience for ANYBODY or ANYTHING!

Now, you egotistical, dumb fuck:

That Sothern writer you maliciously attacked, he had the ability - through his writing and actions - to provide me with hope. And, more importantly, provided me with the right nudge to move forward. To better myself. To look around the world around me and find the blessings in life. And to believe in ME!

Now, you are going to encourage people NOT to read something uplifting that could change their lives for the better? What kind of shit-stain are you to deprive ANYONE of something as wonderful as that? Your puss-like stench of negativity is something I do not tolerate anymore in my life.

Though that good Southern gentleman you purposefully hurt is a better man NOT to stoop to your arena of filth, I'm not, motherfucker.

If I were to see you on the streets, it would give me great pleasure and therapeutic release to kick your fucking teeth down your throat. The sound of your skull cracking would be an inspirational orchestra for me to continue the rapture.

So: FUCK YOU. And fuck all your kind that take shots at people trying so hard to make the world a better, nicer place.

Late June Afternoon

- Taylorsville, North Carolina, 5:13 PM

As some of you may have read earlier, I was one angry writer. I'm not going to apologize for the crude and explicit content I composed in response to a terrible comment on Facebook. There is no room for apologies; when it comes to people who blatantly use words to hurt GOOD people, I have no sympathy for them. But I did not start this column to re-hash what has already been written.

Just before I started writing this, I was engaged in the Second Module of my Media Writing Essentials classes. There are 39 Lessons in the Second Module - the emphasis is on prewriting - and I felt I needed to stop the lesson for the day. It was hard to concentrate on. The material is not hard to comprehend, and I was just in another world... my mind floating in how different and wonderful my days living in the South have been.

Today, in the mail was an envelope from my mother. Inside were two letters I sent to her and my father last year. One, dated May 29, 2021, stated:

I write at the dinner table or standing up at my dresser. The Smith-Corona stays on the dresser, just the right height; in fact, at this very moment, I am moving from one typer to the other. No Joke! I will finish a couple of lines in this letter, then go to the older typewriter to punch out some more lines about wanting to go back to school.

That was last year. My goodness. And I remember writing that letter.

Now, I AM in school. Well, taking online courses, which has to count, right? It's something. Something I didn't have a year ago, and, if I am going to be honest, something I thought was just a pipe dream turned into a reality...

Many things - many wonders - within my life over the past year have profoundly shaped me more than at any other time in my life. Many factors, namely support from people who were brave enough to stick with me or gave me a chance, are the true heroes in my story. People like my wife, parents and sister, readers, and the ever-present and available works from other writers have made it so I can be where I am now.

Oh, I can't forget about the dogs! Talk about therapy. Get yourself a puppy if you need some light in your world.

Our dogs have been a major contributing factor to my happiness. More smiles are spread when looking into the face of an ornery puppy.

My Education

When it comes to the topic of my education, I'm a little sensitive. I barely graduated high school. I spent four years in college but did not finish... however, I continue to "self-educate" myself by reading. Reading whatever cates my fancy. It can be a cookbook, an examination on political unrest in America during the Regan years, literary history, a novel about coming of age children in the 1950s - whatever.

No, I do not have a degree. What I do have is a drive to keep learning.

Reading has been one of the greatest passions of my life. Writing is a close second.

(There is so much WORK to do with certain types of writing, it's not as enjoyable as reading - it's work! The rewards, however, are far better. When you finish a piece of writing, when the pages are stacked up - and YOU ARE DONE... there is no other feeling as complete.)

I'm taking classes again. Online courses, which I feel have made a significant impact on education and furthering education throughout the world. I'm studying Essentials of Media Writing. I feel I need to know more about where New Writing is going, and the various formats and interests of said writing is/will go.

We are no longer living in a world where a writer can type up his words and hand pages into an editor. No, no. We have to learn to edit, promote, pick captivating images to go along with written content, make contacts - all these things taking time away from what essentially needs to be done: WRITE.

Maybe I am too much of an old soul, wanting some of the old ways back. I like typewriters, hardbound dictionaries and thesauruses, landlines, fist fights with other writers... all these things are fading away, their colors losing any shine they had, turning grey into the background of history. Instead of looking a word up in a book, all a person has to do is command a phone to tell them what the hell it means. And where is the fun in that?

Assault charges are filed instead of accepting you got hit.

Where is the music and the chaos and the down-to-earth simplicity of writing a good story?

I guess it no longer exists. Finding time to write ANYTHING is hard enough. Then the designing, the promoting, the changes, the legalities... I can go on and on.

Yet, publishing has never been as easy as it is now, for that I am thankful.

But the dice have changed... cards are flipping over backward - a new type of gamble is on the table. And if you want to succeed, one thing hasn't changed: you have to know the rules to play the game.

A July Day

Today is July 3rd, 2022. Most of the people in the country are getting ready for their BBQs, setting up tents, getting their stockpile of fireworks in order for the celebration of the fourth, and being with family and friends.

The Fourth of July - our Independence Day - will always be a fun holiday for me... but I know what the "bombs bursting in air" means. But this is not a column about the Fourth. No, no.

It's the 3rd. And for me, this is a private holiday; or, more accurately, a tradition. Today, back in 1971, poet/philosopher/lead singer of The Doors, Jim Morrison, died in a Paris apartment. He was 27 years old.

To this day, biographers and memoir writers speculate on the exact nature and circumstances of his death. I have read he died of a common heart attack, heroin overdose, drowned in the bathtub where his body was found - there is even the absurd and perverse theory he faked his own death! That he is still alive somewhere in Jamaica... or Africa... hell, a body of an unknown man was found in a small apartment in Cleveland, Ohio, it was speculated to be Jim Morrison by some of the locals.

Jim Morrison has been a lot of things to me: a teacher, a comforter, an inspiration, and, most importantly, an example of BOTH good and bad things that are available in life. He burned through life too quickly. His greatness is not only by what he wrote but in the tragic sadness in which his life was dealt... and the fatal decisions he made in his pursuit of greatness. How he has overcome such tragedy - such a waste of life - was leaving behind his darkly beautiful work... almost like a puzzle for generations to try to put together.

His dedication to poetry and reading and philosophy and art was something I identified with. His rebel approach and absolute defiance towards authority was very invigorating and inspiring to me... back when I was 20. I'm 38 years old, now. 11 years older than Jim when he passed. As a way to remember and honor him, respectfully, I take the 3rd of every July to dedicate to writing, reading - living! Over the years, I have spent July 3rd in hotel rooms, jail, alone in the comfort of my own apartments and houses...

This year, yes, I will write and remember and celebrate. This year has a little different flavor to it, though. At the end of the night, I'll go to bed with my wife, positive I will live to experience another July... many of them, in fact. Before, they always felt like they were going to be my last.

Sometimes the most rebel thing to do is to say "Fuck you. I ain't going anywhere."

I wish that was a lesson Jim could have learned.

A Writer in the Fields

This writer listened to the crickets. They were invisible, hidden, and camouflaged all through the country night. He wondered why a creature so elusive, so perfectly invisible at night, wanting to be so loud.

Maybe they are lonely at night. They need to talk to one another with their fabulous chorus.

He knew that lonely feeling, especially at night. Earlier, the writer was sitting at his desk, looking past the half-filled page in his typewriter. His work from the night. Half a page. That is nothing to feel bad about. It was even a single-spaced page…

Those words were like all those crickets, too close together, making their own silent concert in the night.

Another Typed Mystery Page Found

The hole showed more crimson than the whiteness of virginal-quality mystery.

Now, looking back at that wonderful bliss of darkness – the terrible, dark part of the night – the hole we look into with no beginnings… with no end, following that available, never-ending wonderment of the unimaginable…

How can we know how many colors are available in the vast stream of the world to narrow it down to the ever-lasting definition of the moment?

As I looked at the picture – looked into dangerous beauty – there was nothing more clear, nothing more evident than waiting… waiting to explore further and further into the image to discover a definition.

Solace Sought Amongst Books

Not sleeping for extended periods is not – I repeat, is not – a skill. It is more like an effective defect. It is an all-purpose, biological trait that is very effective during emergencies.

When Tara and I returned home from the Animal Hospital in Hickory, sending Tara to rest for a few hours did not take much convincing. We had to return to Hickory by eleven in the morning.

Once Tara had made up her mind a visit to the veterinarian was mandatory, I'd made up my mind sleep would be something attained many, many, many hours away. By one-thirty in the afternoon, I had gone over twenty-eight hours without a break. Not once had I so much as put my feet up, or rested my eyes.

My prescription Hemingway sunglasses were much appreciated.

The October afternoon was sunny, sunny-bright and beautiful and warm. Yet, all through the night, farther up the Appalachian Mountains, North Carolina experienced its first snowfall of the season. Leaving in the morning, both the Wife and I were bundled up in heavy jackets when we left the cabin.

As explained on the phone, little Tyke the Terrier was recovering well. He was no longer in need of further medical assistance. June-Bug, on the other hand, was a wreck. She had suffered through several seizures. Her bloodwork showed her liver enzymes were quite high… she was not lifting her head, her tongue hanging out the side of her mouth, and an I.V. was still required to keep her hydrated and help "flush out" her system of, what the Vet called, a "unidentifiable toxin."

June-Bug was transferred from Hickory to the Medicine Man Veterinarian Clinic, located in downtown Morganton. Tara did the driving, with Tyker-Lee resting on her lap. Me and June were stationed in the back seat, her large, white, fluffy frame wrapped in blankets and towels. She was panting fast and heavily, eyes glazed over in drugged-out dumbness.

My heart went out to her.

I know what it's like to have a shotty driver escorting you into an emergency situation when professional medical attention is mandatory.

Once June was checked in, the three of us (Tara, Me, and Tyke) decided to get some much-needed substance – good food and cold drinks!

Absurdity takes many forms. When it comes to Tyke, absurdity is as much a character flaw within him as being hoity-toity. Not only was he obviously feeling better, but he was also prancing like a prince on the dock outside The Town Tavern of Morganton, affectionately dubbed the "River Bar" by Tara. And that is exactly what it is.

Outdoor seating, next to the river, was the main reason we chose that establishment. We have dined there before, and their food is well-filling and tasty (I highly recommend the crab cake sandwich) and proportion perfect.

Now, that little royal, nose-up-in-the-air brat feasted like a king. He even sat on a chair at our table, being hand-fed bits of bacon from my salad, pieces of meat from Tara's meatball sub, and torn-off bits from a toasted hoagie roll dipped in ranch dressing.

Thankfully, not many other patrons were around to witness the luncheon absurdity.

Tara was calling this farce a "celebratory recovery feast from being rescued from the jaws of Death."

Tyker-Lee Locke: 1

Grim Reaper: 0

Good Lord, I thought as my wife went on:

"He was circling the drain… thankfully, he is a survivor against the many traps and pitfalls of Death," she said.

I paid the bill, keeping my mouth shut, curious to hear what other inventive and entertaining proclamation Tara could come up with.

We still had some time before we could retrieve June.

Stopping at Thornwell Books on South Sterling Street seemed the most appropriate place for the three of us to kill an hour. Tara and Tyke could lounge on the second floor, where Tara hoped to get some work done. She was wise enough to pack up and bring her laptop for the trip. Mrs. Locke is one of the fortunate people that can earn money anywhere. That is, as long as she lugs her laptop along and free WIFI is available.

Me?

Hell, it's a bookstore. I can spend an entire day in a bookstore, completely unaware of the rest of the world. A bookstore is a universe in itself, offering its System's cognitive pull to all walking through the entrance.

My mind was tired as an elephant trying to pull its chained leg away from a tree. Sooner or later, the elephant always gives up. That is how the poor beasties are trained to stay…

Anyway, usually, in a bookstore, I rush from one section to the next, trying hard to recall all the titles and all the authors I mean to include in my ever-growing, private library. Today, I just did not have the juice – the existential drive – for an ultra-excited, regular visit.

Finding the Biography/Autobiography section, I took a quick scan through the titles, selected what I thought – what I knew – was the right ticket for the ride I was on, and bought it.

Finding an over-stuffed, dangerously comfortable reading in the True Crime section, I opened my new copy of Eudora Welty's One Writer's Beginnings, willing my eyes to stay focused on the text

(It did not work)

…how long had I been gazing at page IV of the book's Introduction? My eyes, probably shiny, glassy, and unmoving, quickly blinked. My body jerked in a fast, violent shudder.

Where had I been?

I knew where I was and why I was there needed to be answered.

I reasoned to myself: You are tired; nothing is wrong.

My Self felt as though it was slowly returning and resetting for whatever was to come next.

That is how Tyke and Tara found me. No need to mention my brief mind drift. There was still June Bug to collect from the Doc's office. And I needed to stay awake during the ride home. Tara was dragging. And understandably so. Both of us were physically, mentally, and emotionally exhausted, and both of us knew what it was like to be running on desperate fumes.

Not that I have ever worried about her catching some Zs behind the wheel. She is a professional hardass when it comes to grinding away towards a destination.

Her destination was home!

And she was taking all of us back to the cabin.

Her family – all of her pack – are going to be together tonight.

TO BE CONTINUED…

There Is No Place Like Home

Sometime after 3:00 a.m., I crashed. No dreams. No conception of anything until Tara started shaking me awake.

I wondered if I had slept at all...?

It was early. Too early. We were supposed to have June-Bug back at the Medicine Man Veterinarian Clinic by 7:30 a.m. Last night, Doc - Dr. Clarence Dunning - agreed to let June-Bug be discharged so she could spend the night at home. He was gracious and trusting enough, knowing Tara and I would do anything to keep our not-so-little little girl comfortable and safe back at our cabin.

We were armed with medications, medical instruments, and instructions to return to the clinic in the morning.

June-Bug... oh, June, June, June. She was one sad sack of a fluffy lump. With all the stress, being sick, several doses of narcotics, and anxiety, Miss June needed assistance from Tara, Me, and two employees from the clinic to load her limp, heavy body into the Jeep. Her tongue hung out from the side of her mouth like something pale-pink and dead that crawled out of an infected sea. Her eyes were black and shiny slits through mounds of heavy flesh and thick, white fur.

Things were not looking good.

Tara commented: "If she passes away tonight, at least it will be at home."

I silently agreed.

"We are going to have to take shifts. Doc told me she needs some of her medicine every hour."

Every hour, I thought. My goodness, I'm not going to sleep! Again!

"We should keep her on the first floor. In the Boy's room. No sense in me wrestling her up two flights of stairs."

"Good idea. I'll get the bed made up. I'll sleep in the room with her."

It was settled.

And that is what happened.

Tara spent the night in our son's bedroom. I stayed awake, typing things up on my laptop, mostly notes I sketched in my Reporter's notebook, and, on occasion, looking stupidly at the television, not considering any of the images or sounds coming from the screen. I was not even sure why I was staying awake. Maybe it had to do with the fact I had gone long enough to be too tired to fall asleep.

That is a thing.

Trust me.

I did, however, manage to climb up to the third-floor master bedroom after some pitiful hours of piddling around.

Dr. Dunning wanted June back to flush more of her system and to keep observation on her progress... or lack of. Tara told me he thought if she had not improved or was worse off, he would start to worry.

By the time I had gotten dressed, had something to drink, and lugged my way down to the ground level of the cabin, Tara had informed me June seemed better and I should peek in at her before cleaning up for the day.

When I stuck my head into Vincent's room, June had her head up. Her tongue was being used (appropriately panting away, like normal), and she could lift the upper part of her body. Both front paws and legs were working. Half of her body was lifted up. There was strength returning!

"Hell, yes," I mumbled.

We left Tyker-Lee behind on our second trip to Morganton. As written prior, the little shit is fine and full-of-himself-healthy.

I whispered my prayer of Thanks before loading up the Jeep.

June-Bug remained calm on the trip. Most of her body was covered in blankets, that big head of Her's half visible, peeking beneath the stitched, colored patchwork of a blanket whose origins are a mystery to me.

(It probably came from Doc's.)

We - a nice lady employee of the clinic and I - unloaded Ms. June from the backseat, carrying her all the way down a brightly lit hall to the far-end examination room. After she had been checked in, after Tara had her credit card swiped, she and I were on our way back to Thornwell Books.

Now, I am not the type of writer that usually works in public places. I mean, I always have a notebook and pen in my pockets or my shoulder bag, which is filled with all kinds of different things. A great deal of them have nothing to do with writing... however, I need them. At least, I need all the odd and comical little tomfoolery gimcracks lugged around with papers and books, pens, and journals. It is a very common thing for me to be transfixed, my head bent down, looking into a notebook, scribbling notes and ideas.

On this October morning, though, I had my tried-and-true bag filled with electrical cords, a handheld digital recorder, and my laptop.

I was going to be writing - actually writing - in a public retail space, with other people around me shopping, ordering and drinking coffee, talking, and shamelessly kibitzing. It has been a long time since I constructed anything of solid substance in public. Back when I was sixteen and seventeen, I wrote at 24-hour coffee shops, school, and anywhere else there were other people.

Can I still do that, I wondered.

Turns out, I could. And I did. In fact, I had so much fun doing it; the whole three-and-a-half four hours we were at the bookstore flew by. My attention was pulled away from my work when I had to use the restroom or go outside for a smoke.

All through that time, I wrote the piece that was published on October 19th. On the second floor of Thornwell's, I wrote and wrote. Tara was behind me, working on a floor couch. I was lost in the work, enjoying each key I struck, flipping through the pages of my notebook, copying the notes I had taken, and watching the screen fill with words.

Doc called Tara a little after 3 o'clock.

June was ready to be picked up.

Tara went into the clinic; I waited in the Jeep, smoking a cigarette. It was another bright Fall afternoon. Beautiful and warm. I rummaged through the back of the Jeep until I found the yellow cardboard box. Tara had bought that box, filled with little rubber duckies, as a gift for me. She wants me to partake in the Great Jeep Ducky Phenomenon, or Duck. Duck Jeep. It's this silly affair when people participate in placing different types of rubber duckies on strangers' Jeeps, later to be displayed on the dashboard.

I chose a duckie with a medical mask over its beak and a stethoscope dangling from its neck. Doc drives a dark green Jeep Gladiator.

For the second day running, I wedged a ridiculous rubber ducky in the driver's side door handle.

I smiled as I walked back to my own Jeep. It was ridiculous - even asinine - what I had done, but who cares - it made me smile!

When Tara walked out of the office, June was hot on her heels, wagging her big, fuzzy tail away. She even jumped into the backseat.

This was not just luck!

Doc explained she was not completely out of the muck. June-Bug still needed medication, rest, and to stay hydrated.

Both dogs had pulled through. Both, as of this writing, are home. June is still not eating the way she should. I think she simply does not like the fancy food Doc wants her to eat. Tyke... well, Tyke is just Tyke... prancing and proud and entitled.

Tara believes - and I believe it, as well - we would have lost June if we had not taken her in.

This is our Happy Ending.

Thank you, dear Reader, for being there for me to write to. It has been a rough week. We could not have done it without you.

Thanks.

General Roy

He did not know where his last thought went. Like trying and trying to catch something, missing by mere centimeters, there was the brief belief it is there, that you came in contact with it… whatever it may be.

The typewriter was no help. The Writer looked at it… always transfixed in a gaze at the mouth and teeth, stuck in dumb moments, which lead to long staring contests. He hoped it would come back to him. The idea, the flow, the narratives.

Was he going to stand up?

Was he going to remain a warm and fleshy and fragrant lump on his patient's office chair?

An idea – a thought – is a terrible thing to displace. He could spend more tragic time trying to conjure up a lost thought than time spent performing whatever in the hell it was he was going to do.

Is it hunger?

Was he going to get something to eat?

Perhaps he knew what he was going to eat, and now, like everything that was and will become, is gone.

No, no.

It was a cigarette.

That was it.

That's the ticket, he trumpets inside. The Writer was going to set a smoke aflame. Thinking about what to write while the onion-skin thin white paper turns to grey ash.

Reaching for the crushed half-pack of Marlboros, the Writer suddenly knew what to write.

Quickly lighting his smoke, taking a puff, and then touching the glass surface of keys, the room filled up with that special clatter and "ding" of his machine, Roy.

That was his typewriter's name: General Roy.

"Roy, you piece of shit, keep talking to me," he would shout when things were going well. Or: "Good job, my brother. Thank you."

The Writer had been thinking too much. Too much, too much. It was a bad thing he was doing… trying to remember all that he thought.

His friend, however, General Roy, reminded him to keep things simple. To be calm in the face of the most dangerous game a man could play: the Game of Thought.

General Roy smiled at his Writer, his composer, director, and dictator. His dearest of friends, the Writer. The Writer was keeping a secret from him. All the Writer had to do was reach out, touch the black and white pearls of General Roy's smile, and all would be well – for the both of them.

I HOP(e)

It has been a very long time since I have been on a Breakfast Date. A lot was going on this morning: I had an appointment with a General Nurse Practitioner. It was something important I had to do. My medications are important, but deep down, there is that kid in me who HATES doing things like going to the dentist. I despise anything concerning the closest resemblance to a medical procedure.

Then again, what type of person LIKES such things?

Maybe it has to do with the being an "adult" part I despise so much in such affairs. You SHOULD go to the dentist. A person SHOULD go and get checked out to get their medication... but in the primal part of a Hunter-Gatherer Society, is it necessary?

I suffer from Social Anxiety, along with other psychological maladies that make it awkward (sometimes downright frightening) to perform the grown-up role of keeping up with my health.

That.

And I'm an all-out coward!

I had to be in the Practitioner's Office at 8:45 a.m. It started off as a good morning. And it turned into a great day and night. But there was something strange and fun about me this morning. For example, while drinking my morning coffee, reading over my messages, during my time in the bathroom, and brushing my teeth (starting the day), I was listening to Ghetto Sex Rap sung by some very provocative Ebony Music Artists.

Oh, and there was the dancing: I had to dance around the house, listening to this music, making Tara proclaim my insanity.

She agreed to take me out on a Breakfast Date, though, after my appointment.

We agreed on IHOP.

Now, if you have never been to an IHOP in the South, you are being denied one of the greatest pleasures this fine county of Ours has to offer.

As of this writing, there are 1,841 IHOP restaurant locations in the Americas. Owned by Dine Brands Global, the International House of Pancakes (shortened to IHOP in 1973) opened its first doors in the California area in 1958.

My mother was the first person to take me to an IHOP, and she still called it by its original name. This morning was the first time in over two years I graced one of the restaurants with my presence in one of their establishments. The last time I was at an IHOP was terrifying. I was terrified, sad-mad, and sick.

That morning was an odd morning. Odd for me, because of the help I received from some fine employee at an IHOP located in Niles, OH.

It was in October.

That part I remember.

It was RAINING, with SLEET! And it had been raining all night. At the time, I was living on the streets. Two bottles of Coconut Rum. I remember THAT! That was my inventory for the evening.

I was wet.

I was cold.

I was sick.

My cigarettes had long-done been smoked.

But, I had the Rum.

It had gotten to be at the point of the night when everything in small-city Niles was closing and locking its doors for the night. Once again, there was nowhere for me to go. Normally, finding a corner to curl up and protect myself from the wind was easy to find. But this was turning out to be one messed-up night.

For one, I was paranoid.

Stealing bottles of booze in Niles, OH, is NOT the same as jacking some from a department store in Youngstown. In Youngstown, they don't care. Not a one lick!

So, my fear of wandering around Niles, a much more conservative and protective city, homeless and drunk, no cigarettes... if a Police Officer had stumbled upon me, it would have been all bad.

A miracle was discovered that night, though. Something I had not thought about... something that stared at me in the parking lot: a de facto apartment! A place to get out of the rain! A place to drink my poison in peace: a toolshed!

Placed outside of a Lowe's Home Improvement were a dozen (two dozen?) toolsheds. UNLOCKED toolsheds!

(Just what was needed!)

I spend a shivering and cold rainy night in a light grey Little Cottage Company 10-ft x 12-ft Engineered Wood Storage Shed.

The sad part about that was how grateful I was to be in that shed, shirts pulled up & over my head to keep the heat in. I stayed there all night. And I would have stayed longer. But, like most great things, they must end. And I'm not talking about the toolshed. I ran out of booze. Always and always and always: I was running out of booze. Years of my life have been spent waiting for the next drink... fearing I would not have anything to drink.

That morning was no exception.

My hunt for more began.

On this particular morning, however, something different happened. I went into the IHOP next to Lowe's. I had to use the restroom.

Even though I looked like a drowned rat, smelling of sweet and cheap Rum, I thought maybe I could ask for a job application.

It had to have been a little past 7:00 a.m. I waited outside of the restaurant, picking out what half-smoked cigarette butts were salvageable in the ashcan receptacle. IHOP restaurants operate twenty-four hours a day, seven days a week, from 7:00 a.m. to 10:00 p.m.

When I finally made it inside, the warmth from the Heating was amazing.

In my experience, very few people - well, MOST of the people I know - cannot really grasp what being cold is really like. It has become a pet peeve of mine when someone says they are cold.

NO!

NO!

You are just a little uncomfortable.

Being cold HURTS!

I was in the lobby when a short, stout waitress of about 50 took a look at me. At my wet clothes. Wet hair. Dirty. Shivering and shaking from more than just the cold.

"Are you okay, Honey?"

Her eyes never left me.

I was the only person there besides her manager and the cook

But those were eyes I didn't mind looking at me. They were devoid of contempt. There was nothing cruel about them - about her.

There must have been some look of desperation about me because she asked for me to sit so that she would get me a cup of coffee.

"I need to use the restroom," I told her. And I told her sorry that I did not have any money. That if I drank the coffee, I would get sick.

"My name is Joyce. You go to the restroom. But I want you to come right back here when you are done. You understand?"

(I surely did understand. I understood I was dealing with a GOOD person.)

"Yes, ma'am. I'll come back."

"You better! I got something for you when you get back."

At that point, I was past the point of arguing. - I HAD TO GO TO THE BATHROOM!

When I was finished doing what people do in the bathroom, I made my way back to the table. Ms. Joyce was standing next to the table, a large purse on her shoulder.

"You're homeless," she said.

It was not a question.

"Are you an alcoholic?"

I surely am. And I told her so.

She then did something AMAZING. Something SHOCKING. She pulled a fifth of Southern Comfort from her purse. It was over half-full!

"Now, I'm not giving this to you for free. And you don't have money."

Here it comes, I thought. On more than one occasion, I have been asked to do some not-so-nice things for and to some not-so-nice people.

Had I been fooled?

She seemed like a nice Lady. But even nice Ladies tend to their gardens, as well.

"I'll give this to you if you eat something. I have already talked with my manager. You don't have to pay anything. And he wants to talk to you, too. You can pick anything on the menu. Does this sound good to you?"

Boy, did it ever. I couldn't remember the last time I had had a meal of ANY kind! Let alone one in a RESTARAUNT!

And, of course, I was thrilled about the booze.

Wednesday's Web-Cakes and Morticia's Haunted Hot Chocolate - (with generous shots of So-Co) - was my breakfast. And the year was 2019! That is when all this happened. 2019 is when IHOP featured The Addams Family menu.

While eating and drinking away, I asked Ms. Joyce if they were hiring. She was sweet enough to placate me, bringing an application and pen. While filling out the application, David, the Manager, came to the table.

David asked if the food was alright.

"Yes."

He asked if I attended church.

"No."

Well, guess what? Mr. David DID go to a church. He said he called a friend or The Minister - somebody from a church. He had asked said individual if they had any clothes.

They had clothes.

Mr. David asked for a shirt & pants size that his friend would bring them, along with a coat.

Fifteen or so minutes later, I'm in dry clothes. I also had $20. A package of Winston cigarettes. And there was still something left in the bottle to kill.

Some strangers - Strange Angels, as I have come to call people like that - helped me. Really helped me in the best ways that I knew how.

The Strange Angels.

Here, in North Carolina, at their IHOP (MY new IHOP), I met some good people, as well. I'm far from what I was back in Niles. (I sleep in a bed now.) These Southern People, though, with their Southern twang and smiles, make for a fun venture to public establishments.

Our poor waitress spent a good twenty minutes talking to my wife and Me. We, unfortunately, talked a lot about COVID-19. However, Ms. Matt (Yes! That was her name) spoke with laughs...? That's the best way I can describe it. Every sentence had a laugh along with it.

I'm seeing this more and more. Every time I go out to get something, to do something, I am meeting more and more "common" and "special" people that just start TALKING to me. No reason whatsoever.

Tara, my wife, told me I have a way of bringing out the good in people.

I just wanted to shake my head and laugh.

People go up to members of my family all the time, and the same damn thing happens.

Well, here, in the South - the trait is exacerbated.

The Great Sales Horror

Way back when things were very different, I used to sell books. And I mean any kind of book. New. Used. Rare First Editions. I would go to garage sales at Goodwill, and I would check things out online. Buy a book from Amazon.com, knowing it was worth far more than what it was selling for, & I would get it. Once I received the book and gave it a good look-see, I would put it up for auction on a different website. I have gone to Used Bookstores & found Signed First Editions from writers such as Chuck Palahniuk & Norman Mailer!

It blew my mind sometimes.

At one Used Bookstore, I almost got into a fight over a first-edition Hunter S. Thompson book, The Great Shark Hunt. It was priced at $5.00.

Hell, yeah, I thought. What a find!

What happened at check-out was almost epic. Almost ended in the greatest, ugliest tragedy of Used Bookstore history.

"Um, I can't sell you this," the kid behind the counter said.

I knew something was up when he was gawking at the book. And when he started typing away at the computer, I knew I was going to be screwed. I knew this was going to happen. And I was prepared for it.

"Is K___ here? You know, your MANAGER."

He was taken aback. I knew his Manager by name. And hell yes, how could NOT know his name! Scarier still, for the kid, The Manager knew MY name. That bookstore kept accounts for people. And it had a Purchase History file on the account. That's right: ALL of my purchases in more than a two-year span were recorded.

(I even ask for a printout of the bastard. It was a LONG list, my friends.)

So, of course, the Manager, K__, knew me. He rang me up so many times at check-out, asking about my family. The whole pleasant Purchase/Transaction scenario played with perfection between K__ & Me.

And now I'm being told I can't buy a piece of treasure I legitimately found.

K__ showed up on the scene. It did not take him long to figure out something was really wrong. He looked at my face, looked at the book next to the computer...

"What's wrong, Leland? When I shelved that book, I KNEW you were going to be the one to get it. I just KNEW. I almost set it aside for you. So what's the problem?"

"That kid behind you won't sell it to me."

K__ turned to the Sales Associate, a look of anger on his face

"Why the hell not," he snarled.

The kid started talking fast. Telling him about the price, and then pointing to the computer screen.

"It's priced at five bucks, Mr. K___. I expect you to see the problem. It's your problem; someone on the staff priced it wrong. They didn't know a gem when they saw one. Please don't make it my problem."

He didn't say anything for a while. And he didn't look at me. He just looked at the book. Then, he would look at the computer screen, & then back to the book. When he did say something, a section of my mind exploded with rage. Another section melted.

"We can't sell this to you at this price, Leland. Certainly, you, of all people, can understand that."

I was LIVID.

I felt deceived.

This was an OUTRAGE!

"K__," I said as calmly as I could, with all the menace I could muster from my eyes, "the books are sold 'As Is,' and the price is non-negotiable. What the hell, K__? This is ME, a devoted customer. And THIS is what I get!"

He wasn't fazed. Nor did he seem impressed. He knew that I knew the worth of the book. He probably also knew that if I were in his shoes, there was no way I was going to let that book go for five measly dollars.

I turned rude really quickly.

"So what the F#@K you gonna do?"

"Let's work on this, Leland. I'll make you a deal: I'm going to put this in the Rare Book display area. I'm going to give it a price of $95.00. You can buy it now at that price."

My breathing was not doing well. I could feel the muscles in my jaw starting to clench, tighten, relax, and then begin again, distorting my facial features.

He saw it right away. I was going to explode.

"Now," he began his pitch, "if this does not sell in two weeks, I'll pull it from the shelf. I'll call you & can buy it for, say, $35.00. Is that fair?"

The bastard had me. $35.00! I would have paid for that at that moment. If it was $35.00 when I saw it on the shelf, I would have bought it.

The only thing: what if someone bought it? What if it sold that very day, which, I knew, as soon as I was out of the store, that's what was going to happen.

"What if it sells, K__? What then? I just swallow the fact you F@#KED one of your best customers! Is that the type of S#$T you pull?"

He was quick. And Smart.

"If it sells, Leland, you can have 75% off your ENTIRE next purchase."

"I want 75% off the entire total amount on my next visit. Get me? Not ONE book. ALL OF THEM!"

With that, he hesitated.

"All right, Leland."

He went to his office. I stood glaring at the kid. He did not want to look at me. The Great Shark Hunt lay two feet away from - a monument/artifact of significant, personal worth. I wasn't going to sell it. Not at all. It would stay in my Private Collection if I could wait two weeks... if it did not sell... if I did not break & buy it for $95.00.

K__ returned with the store's business card. Written on the back was the agreement for my 75% off. It was initialed. And he said if I had any trouble, to call him. He wrote his cell phone number down, too.

I was not happy when I left. And I tortured myself for the next two weeks, looking the book up online. Looking at pictures of it. Rereading the trade paperback copy, I'd had on my shelves for years.

I got the phone call.

It was on a Wednesday.

"Leland! It's K__ from Used Books. Come and get your book. I'm putting it behind the counter for you."

"I'm on my way."

I walked to the store. I was happy as hell. I started my First Edition Hunter S. Thompson Collection earlier this year. This buy would put my collection at eight titles. All of them had wonderful dust jackets. It was going well.

K__ was the one to ring me up. After he handed the book to me, he said, very sternly, "Give me that card I gave you." I smiled. He knew me well, too well if you ask me.

I fumbled around in my wallet, handed it over to him, and then he did something that made BOTH of us smile: he took a black Sharpie, crossed out the agreement, handed it back to me, saying:

"That should make a good bookmark."

As far as I know, that card stayed in that book for as long as I had it.

I do not wish to share with you the terrible fate of my book collection. It makes me shudder thinking about it.

I feel I am making a slow comeback, though.

I'm two titles back in the game...

and counting.

My Literary Treasure

Had to go through my email today. Had to clean the sucker. And to see if there was any mail from someone, a certain someone I don't want to get messages from.

In the process of deleting all the credit card offers, offers to lower my credit card payments, more offers for more credit, and, of course, the pitches for great car loans, I came across something AMAZING. It was in the SENT files. It was a large SENT file to an editor I had contacted back in 2013.

Attached to the file were a dozen or more Documents I had sent to him. - My writing. I sent it to him with a desperate plea to see if he could do anything with the work. Included with the Short Stories & Poems was my first novel - and it had been edited! That's right. With a billion red lines slashing through my half-crazed sentences, & red words & letters going over & covering one another like some grotesque Marquis de Sade orgy, there was my novel. Edited to a disfigured point of unrecognizable obscurity.

But I have a copy of it!

All 255 pages of her are here, with me... well, with the Electronic & Digital Gods of online heaven.

It was something I thought I would never see again! Mangled with improvement? Sure. I can call it that. But the point being, I have her.

I used to pages and pages and pages of writing. All of it I thought I would never see or read again. The other work, the Poems & such, are an added delight. But they are just sprinkles on the sundae. My delicious triple scoop is that novel. I spent the better part of two years working on it.

And, oh, I worked hell on it. At that time in my life, I felt writing that manuscript was the most important thing I had to do. Forget college. That was secondary. My job at the bookstore was entertainment.

No, no.

It was the Novel that had me.

In all honesty, it's not that good.

But it was the first big thing I had ever written. One of the first things I did that I was dedicated to.

Whenever the printer is figured out here on the Compound, I'll print it. Having a physical copy seems important, even if it's just to gather dust in a drawer.

Made in the USA
Monee, IL
02 November 2026

67385376R00142